W9-APV-794

California Natural History Guides: 37

Geology of the Sierra Nevada

by Mary Hill

Maps by Adrienne E. Morgan

Drawings by Alex Eng and others

UNIVERSITY OF CALIFORNIA PRESS
BERKELEY · LOS ANGELES · LONDON

California Natural History Guides
Phyllis M. Faber and Bruce M. Pavlik, General Editors

University of California Press
Berkeley and Los Angeles, California
University of California Press, Ltd.
London, England

ISBN: 0-520-02698-5 (alk. paper)
Library of Congress Catalog Card Number: 73-93053
Printed in the United States of America

11 10 9 8 7 6 5 4 3

15 14 13 12 11 10 09 08

The paper used in this publication meets the minimum requirements of ANSI/NISO Z39.48-1992 (R 1997) (*Permanence of Paper*). ♾

SOURCES AND ACKNOWLEDGMENTS

Maps 1-14, index map, and figures 1, 2, 4, 11, 13, 27, 38, and 49 were drawn by Adrienne E. Morgan. Figures 6, 7, 8, 9, 12, 14, 23, 24, 29, 30, 31, 33, 40, 41, 55, and 58 are pen-and-ink sketches by Alex Eng.

Figures 15 and 25 were redrawn by Ed Foster; Mr. Foster also drew figures 17, 21, 22, and 32. Ron Morgan drew figures 39, 50, 51, 53, 56, and 57. All of Mr. Morgan's were previously published in *California Geology*. Both Mr. Foster's and Mr. Morgan's work is by courtesy of the California Division of Mines and Geology.

Figures 10, 15, 20, 44, 45, 47, and 48 are from *Geologic history of the Yosemite Valley,* by François Matthes, published by the U.S. Geological Survey as its Professional Paper 160 in 1930. The U.S. Geological Survey also prepared figures 34, 35, and 54.

Figure 18 is a wood engraving by R. Swain Gifford, from *Picturesque America* (1895) part 18, p. 139. Figure 19 is a wood engraving by Thomas Moran, published in *Century Magazine,* 1890. Figure 46 is an etching by J. A. Fraser, also from *Century Magazine,* 1890. Figure 42 is also from that source. Figure 26 is from *Underground,* by Thomas W. Knox (1873). Figure 28 is a drawing by Garniss H. Curtis, first published in University of California Publications in Geological Sciences (1954), vol. 29, no. 9. Figures 36 and 37 are by W. C. Putman, first published in *The Geographical Review* (1938). Figure 43 is by W. M. Davis, first published in the *Scottish Geographical Magazine* (1906).

The quotation from John Muir, pp. 124-126, is from *The Mountains of California* (New York: The Century Company, 1894). Willard D. Johnson's account of his descent into a bergschrund (pp. 131-133) is from "Maturity in alpine glacial erosion," published in the *Journal of Geology* (1904): 573-574. Muir's account of the earthquake of 1872 is from *The Yosemite* (New York: The Century Company, 1912), pp. 76-86. "The mountains grow, unnoticed" (Chapter 11) is the title of a poem by Emily Dickinson.

Figure 3 was modified from A. Holmes "A revised geological time scale," published in *Transactions* of the Edinburgh Geological Society (1959), vol. 17, pp. 183-216. Figures 13, 27, and 49 used as a base *State of California preliminary fault and geologic map,* scale 1:750,000, by Charles W. Jennings (California Division of Mines and Geology Preliminary Report 13, 1974). Figure 16 was modified from N. L. Bowen's *Evolution of the igneous rocks* (Princeton University Press, 1928). Figure 17 was taken – with changes – from *Chronology of emplacement of Mesozoic batholithic complexes in California and western Nevada,* issued as U.S. Geological Survey Professional Paper 623 (1970). Figure 31 was redrawn from a scanning electron micrograph made by G. H. Heiken, published in the Geological Society of America *Bulletin* (January, 1972), vol. 83, p. 96.

Table 8 was compiled from *Guidebook to Quaternary geology of east-central Sierra Nevada,* by Michael F. Sheridan (published by the author, 1971), p. 5; from *Glacial and Pleistocene history of the Mammoth Lakes Sierra, California, a geologic guidebook,* by Robert R. Curry (University of Montana Department of Geology, 1971) Geological Series Publication Number 11, p. 6; and from *Glossary of geology* (American Geological Institute, 1972). Table 9 was modified from "Glacier trails of California," by John L. Burnett, *Mineral Information Service* (1964), vol. 17, no. 3.

In the section of colored plates, the photo of Lembert Dome is by Salem J. Rice; the photos of the top of the Postpile and eroded battlements are by Susan Moyer.

My thanks to Elisabeth L. Egenhoff, Gordon B. Oakeshott, John L. Burnett, and Frederick A. F. Berry, whose careful reading of the text prompted many improvements in its technical quality, and particularly to Susan Moyer, who tested, criticized, and considered every word in several drafts.

CONTENTS

Chapter 1

SIERRA NEVADA THROUGH THE AGES

Although to us, whose days are brief, the Sierra Nevada seems immensely old (part of the "everlasting hills"), it is not, for it reveals but a tenth of the earth's tumultuous four-and-a-half-billion-year history. The record written in Sierran rocks starts hundreds of millions of years ago, in the middle of things. It opens in the depths of seas we will never sail, pushed aside by rising mountains we will never climb, drained by rivers we will never swim rushing through tropical forests inhabited by animals we will never see and birds we will never hear. It is a tale of steaming volcanoes and chilling glacial ice; of quiet, warm, shallow seas and sudden earthquakes. This book is an account of those events, and of how we know that they took place.

The story begins with an explanation of geologic time in Chapter 2, for time is the essence of geology. Because new methods of measuring time have given us a chronology of the earth in actual years, the words describing relative time (the "geologic time scale") are rarely used in this book.

Chapter 3 describes the Sierra today: where its rivers run, how it is shaped, what its climate is like. Today, the rain and snow, the rivers and streams, the cold and bitter winds are at work altering the mountains, wearing them down, carrying them away – even as we describe them, they are changing.

The life story of the Sierra Nevada begins in Chapter 4, condensing the events of the first 310 million years into a few pages. Our knowledge of those far-off days is scanty; the first record in the Sierra is found in rocks 440 million years old. From then, until about 130 million years ago, the area was beneath the sea. About 210 million years ago, the Sierra as we know it began to take shape: deep within the earth, bodies of molten granite began to cool into rock – a process that took 130 million years to complete.

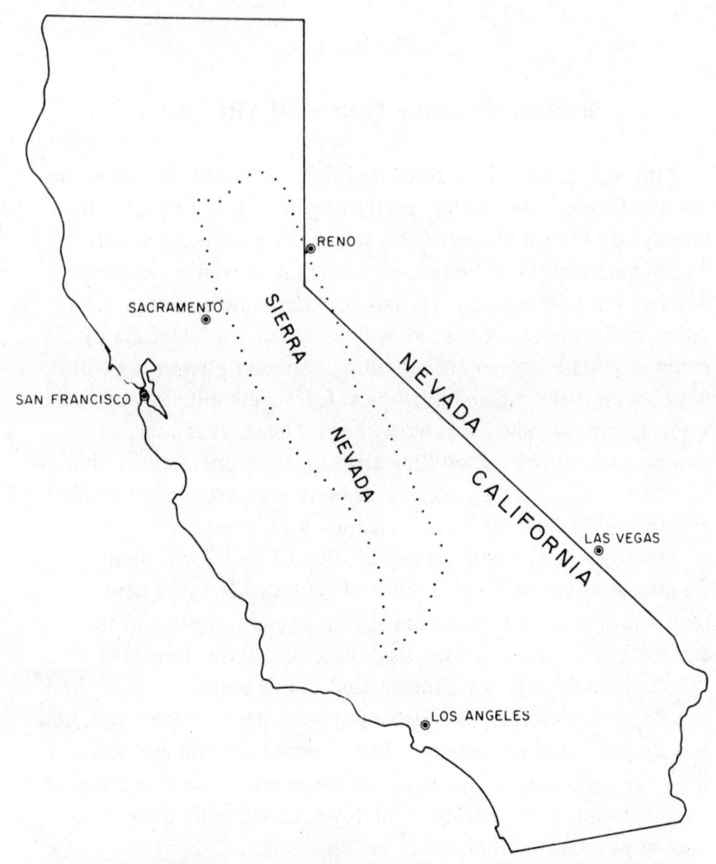

Fig. 1. Location of the Sierra Nevada. The range is geologically related to mountains in the Peninsular Ranges of southern California and to the Klamath Mountains to the north.

The formation of Sierran granite is explained in Chapter 5 – how the hot rock, as it rose toward the surface (but not *to* the surface), disrupted the layers of sediments that had been laid down at the bottom of the sea, pushing them aside, melting some, changing others. Today, these old layers are to be seen standing on end in the foothills. In the high Sierra, blobs of the old rock, caught up in the younger granite, are left near the tops of the high peaks – remnants of a vanished sea that have not been eroded away. Valuable mineral deposits formed at some of the junctions of the old rocks and granite. Gold and other metals were left in veins and fissures as the last of the granite cooled. Chapter 6 notes of some of the mines that have been dug in searching for these minerals.

While the last of the granitic bodies was cooling, and while the gold-bearing quartz veins were forming beneath the surface of the earth, the Sierra was pushed upward, out of the sea. The forces of erosion then began to attack the new mountain range, as Chapter 7 tells, tearing rock from the mountainside to wash down rivers into a shallow, subtropical, lagoon-margined sea. Sierran rivers – not all where they are today – ran in broad valleys flanked by hills rising gently eastward. The crest at that time probably was not much more than 3000 feet high, with only a few isolated peaks rising higher. Altogether, while erosion of the range progressed, more than nine vertical miles of rock was eroded in the space of 25 million years, to be deposited in the sea to form the foundation of today's Great Valley.

But low mountains and quiet seas did not persist. Chapter 8 is the story of the volcanoes that began to erupt about 30 million years ago. Violent explosions threw a blanket of hot ash over the northern Sierra. The ash was followed by volcanic mud flows that almost obliterated the shape of the Sierra as it had been. Before the volcanic fortissimo had begun to diminish, the whole northern half of the range was buried under a sea of hot volcanic mud, leaving only a few projecting peaks.

Shortly after this, the Sierra commenced to rise rapidly. The earth grew colder; glaciers formed, changing the shape of the peaks and valleys. The years of the great glaciers – sculptors with ice and snow – are the subject of Chapter 9.

By whatever means the Sierra was lifted up high enough for glaciers to form, faulting and accompanying earthquakes surely had a part, as may be seen by the events described in Chapter 10.

The chronology of the Sierra has not been easy to determine. It has taken hundreds of man-years of work to bring us to our present understanding of its complex history. Even so, we are looking through a glass darkly, for there are so many unresolved problems that our ideas of Sierran history and origin will surely change drastically. Chapter 11 gives one view of how the range might have been created – a view that may well be partly out of date even as you read this.

For those who wish to see the evidence for themselves, some chapters have listings of where to see examples of the rocks and features on which these ideas are based. They are keyed to Maps 1-14, on pp. 20-37. Some of the lists can also be used on conjunction with the rock key beginning on p. 203, or with the list of rock names on p. 194.

The Sierra is a dynamic mountain range, still in the full vigor of its youth. But we must make an effort to realize this, since little of the drama of its creation is forced upon us. The seas are long gone; the volcanoes are presently quiet; even the earthquakes are a sometime thing that the mountains may not feel strongly in our lifetime.

It is through eyes opened by geology that we can envision the turbulent history of the high peaks. It is by observation of broad vistas as well as microscopic details that we can look backward into time to see for ourselves the progress of Sierran yesterdays.

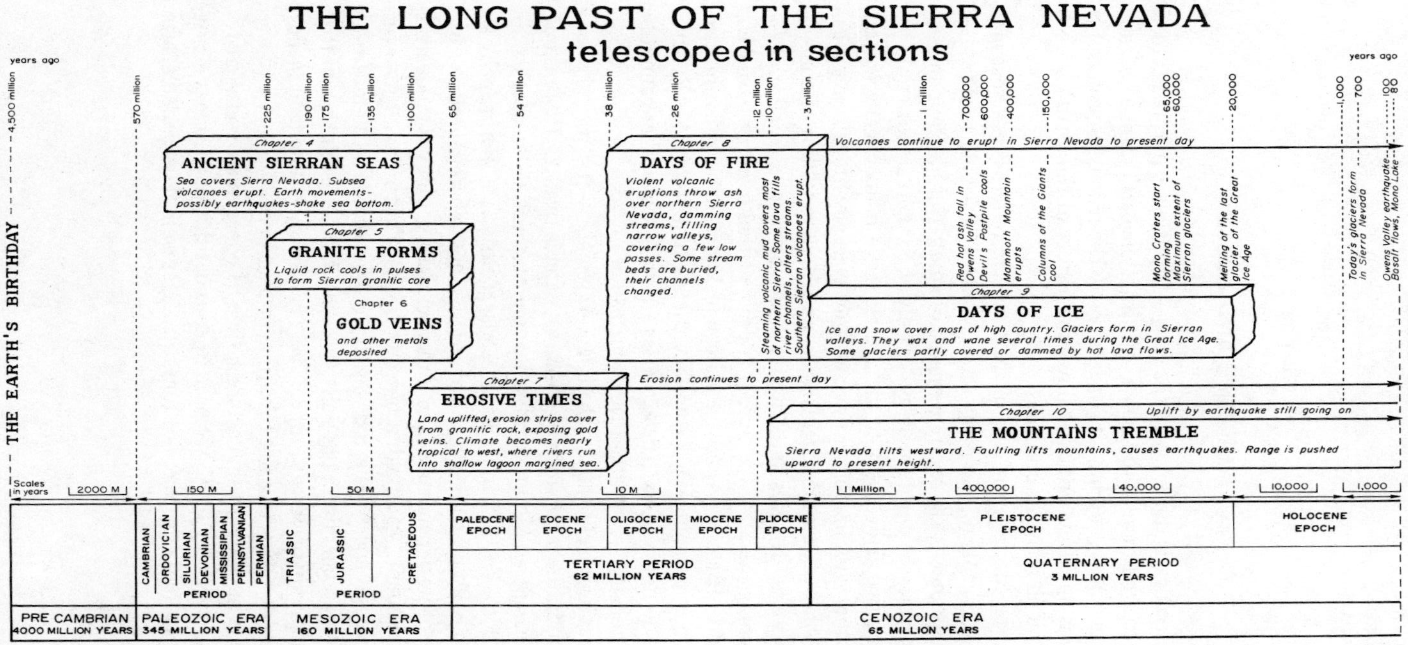

THE LONG PAST OF THE SIERRA NEVADA
telescoped in sections

years ago

4,500 million — THE EARTH'S BIRTHDAY
570 million
225 million
190 million
175 million
135 million
100 million
65 million
54 million
38 million
26 million
12 million
10 million
3 million
1 million
700,000 — Red hot ash fall in Owens Valley
600,000 — Devil's Postpile cools
400,000 — Mammoth Mountain erupts
150,000 — Columns of the Giants cool
65,000 — Mono Craters start forming
60,000 — Maximum extent of Sierran glaciers
20,000 — Melting of the last glacier of the Great Ice Age
1,000
700 — Today's glaciers form in Sierra Nevada
100 — Owens Valley earthquake
80 — Basalt flows, Mono Lake

years ago — THE FUTURE

Chapter 4
ANCIENT SIERRAN SEAS
Sea covers Sierra Nevada. Subsea volcanoes erupt. Earth movements-possibly earthquakes-shake sea bottom.

Chapter 5
GRANITE FORMS
Liquid rock cools in pulses to form Sierran granitic core

Chapter 6
GOLD VEINS
and other metals deposited

Chapter 7
EROSIVE TIMES
Land uplifted, erosion strips cover from granitic rock, exposing gold veins. Climate becomes nearly tropical to west, where rivers run into shallow lagoon margined sea.

Erosion continues to present day

Chapter 8
DAYS OF FIRE
Violent volcanic eruptions throw ash over northern Sierra Nevada, damming streams, filling narrow valleys, covering a few low passes. Some stream beds are buried, their channels changed.

Steaming volcanic mud covers most of northern Sierra. Some lava fills river channels, alters streams. Southern Sierran volcanoes erupt.

Volcanoes continue to erupt in Sierra Nevada to present day

Chapter 9
DAYS OF ICE
Ice and snow cover most of high country. Glaciers form in Sierran valleys. They wax and wane several times during the Great Ice Age. Some glaciers partly covered or dammed by hot lava flows.

Chapter 10
THE MOUNTAINS TREMBLE
Sierra Nevada tilts westward. Faulting lifts mountains, causes earthquakes. Range is pushed upward to present height.

Uplift by earthquake still going on

Scales in years: 2000 M | 150 M | 50 M | 10 M | 1 Million | 400,000 | 40,000 | 10,000 | 1,000

PRE CAMBRIAN — 4000 MILLION YEARS

PALEOZOIC ERA — 345 MILLION YEARS
CAMBRIAN, ORDOVICIAN, SILURIAN, DEVONIAN, MISSISSIPIAN, PENNSYLVANIAN, PERMIAN — PERIOD

MESOZOIC ERA — 160 MILLION YEARS
TRIASSIC, JURASSIC, CRETACEOUS — PERIOD

CENOZOIC ERA — 65 MILLION YEARS
TERTIARY PERIOD — 62 MILLION YEARS: PALEOCENE EPOCH, EOCENE EPOCH, OLIGOCENE EPOCH, MIOCENE EPOCH, PLIOCENE EPOCH
QUATERNARY PERIOD — 3 MILLION YEARS: PLEISTOCENE EPOCH, HOLOCENE EPOCH

Fig. 2.

Chapter 2

OF TIME AND ROCKS

Geology is the study of the earth. Its ultimate purpose is to discover all there is to know about what the earth is made of, how it is arranged, and how it got that way. The reconstruction of history is one of its principal aims: the history of the earth itself and of all its living creatures.

Even though man's use of the earth's resources is as old as man himself, geology, like many sciences, dates back only to the Renaissance. As a separate branch of study, it is much younger than that; the word "geology" is only about two centuries old.

A geologist works with the techniques of many other sciences, together with liberal borrowings from everyday life, a heavy measure of logic, a wild imagination, and a large sprinkling of common sense. He uses inductive reasoning, as does a detective, searching the earth for clues that can be fitted together to make a reasonable story. Almost never can he know for sure that his reasoning is correct. His is always circumstantial evidence. His culprits do not confess. His witnesses are mute stones that cannot be interrogated – yet they reveal much.

Geology lacks the very hallmark of science – it can rarely be tested in the laboratory to obtain repeatable results. There is but one earth; we cannot create a new one to check theories of its creation. For this and other reasons, geology is still principally a science of observation and deduction – not experimentation. In spite of its dependence upon other sciences, geology is uniquely independent. Although it borrows freely, it has a tool of its own that sets it apart. This is a way of thinking that is different from the way a chemist, a physicist, or even a biologist thinks, and that is deeply involved with time. There is scarcely a statement made in geology that does not take time into account.

It is now urgent that geologic time be consciously considered in our everyday lives. We have to consider the long reaches – not only yesterday's time, but tomorrow's, as well. For example, after

the murderous Battle of Hastings in England, in 1066, men and horses were hastily swept into a common grave. Now, after 900 years, the water supply of the town of Hastings is being contaminated by them. Knowledge of the geology of Hastings and a consideration of geologic time could have predicted this.

And knowledge of geology must be used to find ways to dispose of the nuclear waste now piling up in nations of the world. It is extremely dangerous, and must be disposed of in such a way that the people of Earth will be protected from it for a quarter of a million years, until its radioactive clock has run down.

For these, and a host of other reasons, geology has become imperative.

Recently, we have found means to tell the age of a fossil, rock, or past event in actual years – partly by using this same radioactive clock.

Of the many methods used to count the years past, radioactivity is the most useful. This method involves calculating the rate of decay of radioactive elements and the measurement of minute quantities of isotopes of certain elements. For example, the uranium-lead method assumes that uranium, at the time the uranium-bearing mineral crystallized, contained no lead. As time passes, the uranium decays through a series of intermediate products into lead. Since the rate at which this decay takes place is constant, the age of the uranium itself, and therefore the rock containing it, can be calculated.

Besides the uranium-lead techniques, the decay rates of carbon-14, lead-thorium, potassium-argon, rubidium-strontium, and helium are also used. Dating by potassium-argon has been particularly useful in deciphering the story of Sierran granite, and carbon-14 has helped to clarify the last few thousand years of the great Ice Age.

Like uranium, carbon-14 has a steady radioactive decay rate, but a considerably faster one. One of the carbon isotopes, known as carbon-14, is obtained from the atmosphere by all organisms during their lifetimes. When a living thing dies, and is buried, carbon-14 no longer enters its system, and whatever carbon-14 was in its body commences to decay to another carbon isotope.

Wood is a particularly good subject for carbon-14 analysis, for even if it is burned, it can still be analyzed.

Carbon-14 has recently been used in conjunction with another dating method – simple counting – to give a cross-calibration of the accuracy of the carbon-14 method. What is counted in this unusual check are the rings of the rare and ancient bristlecone pine. Since some of these trees have been producing annual growth rings for more than 4000 years – they are the world's oldest living things – by counting the rings backward, it is possible to know which ring was produced in what year. Each ring is then carefully scraped off separately and burned. The ashes are analyzed for radioactive carbon-14. The results, then, are a check on the carbon-14 method: since the exact year is known by counting the tree rings, much of the possible doubt is removed from the carbon-14 technique (at least as far back as 4000 years) by knowing which carbon-14 readings are associated with which calendar year.

Another method of arriving at dates by counting is to enumerate "varves" – the dark and light layers deposited annually on the bottom of many lakes. Each year a black layer and a tan layer mark winter and summer, and by counting each pair, one can tick off the passing years.

Another way of measuring time involves radiation damage to crystal structures. Still another requires assessing the length of time certain chemical reactions take, such as the breakdown of amino acids.

None of these methods is foolproof. The last two and the methods using radioactivity depend upon careful sampling and laboratory work, as well as upon fundamental assumptions that may not be wholly correct.

In this book the story of the Sierra is told in years, but one must recognize that the dates given are subject to change as new information is discovered.

In the past centuries geologists had no way of counting those years, so they used what tools they had – fossils – and based their ideas of the passage of time on the life and death of animals and plants. With fossils as the key, scientists developed the geologic time scale (figure 3), based on organic evolution. Parts of it

RELATIVE GEOLOGIC TIME				ATOMIC TIME (in millions of years)
Era	Period		Epoch	
Cenozoic	Quaternary		Holocene	
			Pleistocene	2–3
	Tertiary		Pliocene	12
			Miocene	26
			Oligocene	37–38
			Eocene	53–54
			Paleocene	65
Mesozoic	Cretaceous		Late Early	136
	Jurassic		Late Middle Early	190–195
	Triassic		Late Middle Early	225
Paleozoic	Permian		Late Early	280
	Carboniferous Systems	Pennsylvanian	Late Middle Early	
		Mississippian	Late Early	345
	Devonian		Late Middle Early	395
	Silurian		Late Middle Early	430–440
	Ordovician		Late Middle Early	500
	Cambrian		Late Middle Early	570
Precambrian				3,600+

Fig. 3. The geologic time scale.

date back to the seventeenth century, when the intellectual revolution began to stir men's minds to speculate scientifically about the earth. Its originators were still tied to the Renaissance, and as they developed this time scale, still one of the geologist's most important working tools, they used Greek and Latin roots. Later other words from Europe and America were added.

The time scale itself is built on the evolution of life – the idea that animals and plants developed from simple forms into more complex ones. It is a story that fits together widely scattered fragments gathered throughout the world. Yet the fit depends upon just a few geologic principles that may seem almost self evident to us now, but were startling and revolutionary when first suggested.

One principle is based upon the observation that many rocks are in layers, heaped above one another like layers of a cake. Therefore, it seems reasonable that the lowest layer was laid down first, with the others piled in succession on top of it. If this is true, one can reason a step further: if there are rocks that now stand on end, or are twisted, bent, broken, and gnarled, but otherwise resemble the rocks of the horizontal layers, then they, too, were probably horizontal when first laid down. The forces of earth itself have moved, altered, and rearranged them. This was an astonishingly long intellectual leap, reasonable though it may seem. Earlier ideas either attributed the twisted rocks to the forces of evil or considered them to have always been that way.

Once the idea of change was accepted, it was possible to understand that animals and plants change also. It was possible to show that there is a definite progression or "evolution" of the form of organisms. If one couples this idea – that animals and plants change through time – with the idea that the layers of rock in which the fossilized remains of plants and animals are found were deposited horizontally, one can derive another important geologic principle: if two rock layers contain groups of organisms that are similar, the layers themselves were originally formed about the same time, regardless of their present configuration or geographical location. This is the principle of correlation. Sometimes the results of fossil correlation can be checked by determining the actual age in years of rock layers.

It is by piecing together clues from rocks in various areas that our idea of the past is developed. Most rocks have contemporaries, both nearby and in distant parts. They are not necessarily the same kind of rocks; after all, rivers run, volcanoes erupt, and tides move the sea all at the same time, and each leaves a different mark – a different piece of the story. By matching fossils and using inductive reasoning, it is possible to match – to correlate – rocks of the same age over wide distances. In this way, a picture of the earth through time gradually emerges, allowing us to see it as it was at various times in the past. For example, groups of land animals alive in eastern South America and western Africa 200 million years ago were comprised of the same animal species. This, together with other evidence, suggests that those continents were once connected and have since pulled apart.

That some rocks were collected in water has been recognized throughout much of human history, but only in the course of systematizing geology have these and other rocks been carefully studied and classified. Today, rocks are generally sorted by geologists into three classes based on their origin: sedimentary, igneous, and metamorphic.

The sedimentary rocks include most of the layered rocks. It is easy to see how the name is derived: if one shakes a glass of muddy water, the mud – sediment – gradually falls to the bottom to form a layer, just as layers of sedimentary rocks were formed at the bottoms of lakes, rivers, and the sea. Some, but not many, of these rocks were not water laid – ancient sand dunes now turned to stone, for example.

The igneous rocks, named from the same root as the word "ignite," were once molten – the lavas, poured out on top of the ground, and such rock as granite, cooled in the deep recesses of the earth without bursting onto the surface. The great mass of the earth itself is an igneous rock.

The metamorphic rocks are those that through the earth's unremitting restlessness have been changed from their original form into something else – something that may be quite different. Limestone may be altered to marble, peat to coal, or shale to slate. Some of the changes are obvious, but others require careful reasoning and the help of many other branches of science to

decipher. Most metamorphic rocks have been formed by the intense heat, high pressure, and chemical changes involved in mountain building.

Geologists divide each of the three classes of rocks into many smaller units, depending upon their composition and origin. For example, the sedimentary group is divided into those that were accumulated in water or on land solely from fragments of other rocks, those that were made from the remains of dead animals and plants or were in some other way derived from them, and those that were chemically precipitated from sea water or other natural solutions on the earth's surface. Coral, peat, and some beds of limestone were once living matter, now turned to rock. Beds of rock salt, chert, potash, and some beds of limestone, in contrast, crystallized out of earth solutions without life processes intervening. Thus, we could refer to "organic" sedimentary rocks, that were derived from once-living things, and "inorganic" sedimentary rocks, that were not.

Sedimentary rocks are also given names according to the size and arrangement of the pieces in them. These are divisions most people know, even though they may not be aware of the mathematical lines of division that engineers and geologists have devised: "mudstone" is mud, now become stone; "sandstone" is just what the name implies; "conglomerate" is a rock consolidated from gravel mixed with sand and mud.

The igneous rock group also has some familiar names in it, although some of them have been redefined by scientists. A coarse-grained, salt-and-pepper crystalline rock most people know as "granite;" "lava" is a common term; "obsidian" (volcanic glass) and "pumice" are other types of readily recognized igneous rocks.

Through the years, geologists specializing in the study of igneous rocks have set up elaborate groupings of them, with complicated rules for naming. To apply these more sophisticated names, laboratory methods not available to most people are required. To become adept at separating the various types of igneous rocks, even on a fairly simple basis, it is necessary to be able to recognize some minerals. Minerals are natural chemical elements or compounds. Most of them are crystalline solids.

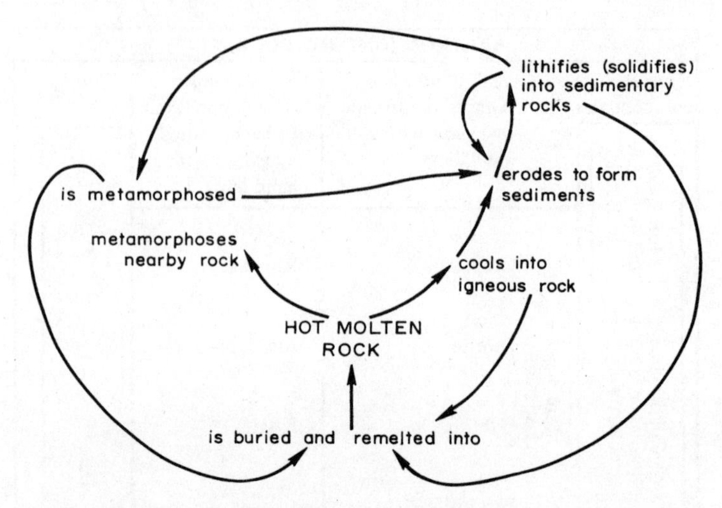

Fig. 4. The rock cycle. The three classes of rock – igneous, sedimentary, and metamorphic – grade into one another, and are, through time, transformed one into another. Hot fluid magma, for example, may cool into igneous rock, either on the surface of the earth as lava, or beneath it, perhaps as granite. The lava flow or the granite may be worn to fragments by wind and weather, to accumulate in sedimentary layers, which lithify into rock. The sedimentary rock may itself be eroded into fragments ("reworked") to become new layers of sediment, and again lithify into sedimentary rock. Perhaps it is then buried deeply; it may be metamorphosed or remelted. Metamorphic rock may, in turn, be eroded to form sediments or remelted to start the cycle over.

They are the ingredients of rocks. Some rocks are made up of one mineral only; others have a dozen or more. California was the first state to have an official state mineral – gold. It also has an official state rock – serpentine, which may be made up of several related minerals.

Although knowledge of many minerals is helpful, recognizing a few will serve to identify the igneous rocks in the key beginning on p. 78. You should learn to know these few on sight: quartz, feldspar, mica, amphibole, and pyroxene. None is truly a single mineral species. They are all families of minerals. If you can recognize the family, you can use the charts in this book. Rocks may contain a wide variety of minerals, but more than three-

COMMON IGNEOUS ROCKS				
Mineral components	Plutonic (Grains identifiable by eye or with a hand lens)		Volcanic (Grains too small to be identified by eye or with a hand lens)	Color
↑ More quartz and orthoclase feldspar	Granite	Porphyry	Rhyolite	↑ Light
↓ More dark minerals (pyroxene, amphibole, olivine)	↓ More plagioclase feldspar: Diorite		Andesite	
	Gabbro		Basalt	
	Peridotite			↓ Dark

Fig. 5.

quarters of the ingredients in all three classes of rocks are members of but two mineral families: quartz and feldspar. If you can recognize these, you have already gone a long way toward learning to distinguish rocks from one another.

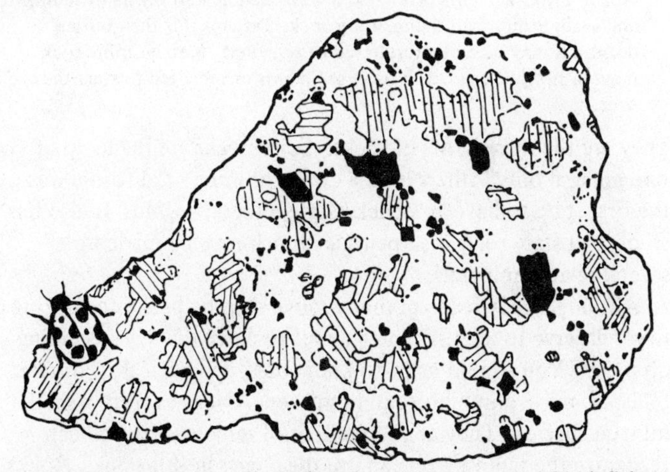

Fig. 6. Granite, showing typical salt-and-pepper appearance. Black flakes are biotite mica; mineral grains marked by lines are feldspar; white areas are quartz.

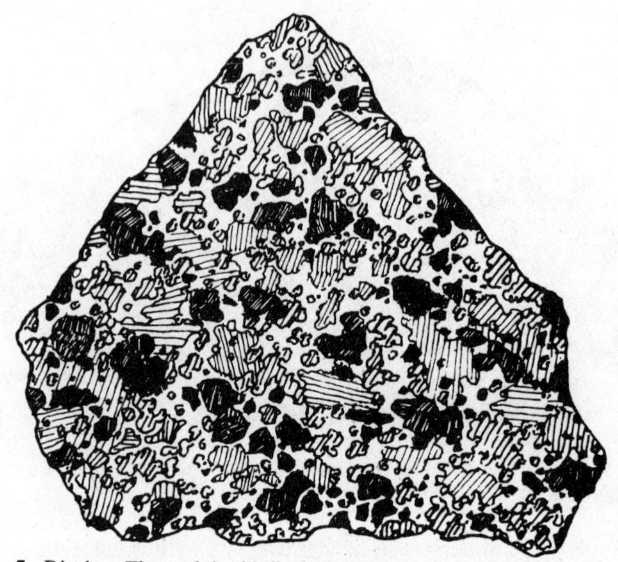

Fig. 7. Diorite. The rock is chiefly feldspar, with dark minerals more in evidence than in granite. Very little, if any, quartz is to be seen.

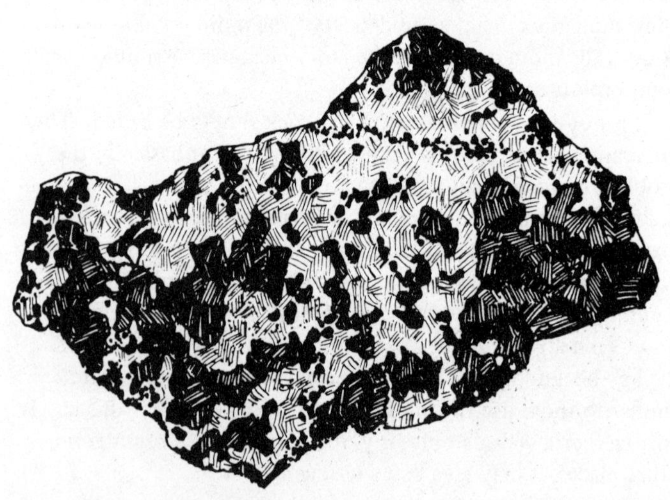

Fig. 8. Gabbro. The rock consists almost entirely of feldspar (plagioclase variety) and dark minerals, in almost equal amounts.

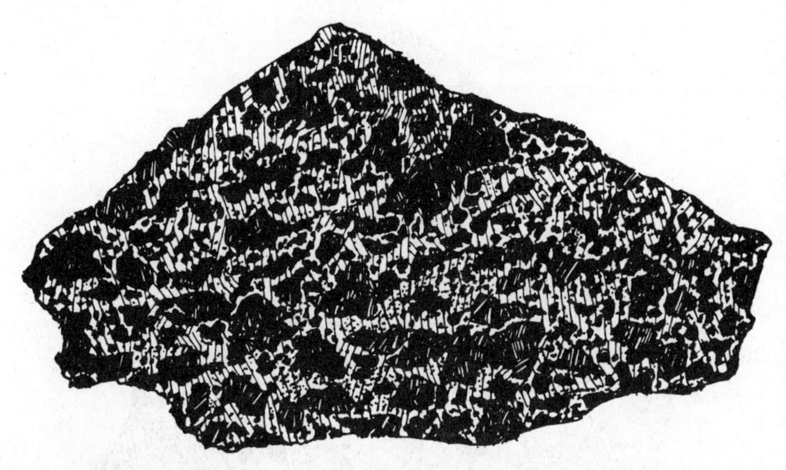

Fig. 9. Peridotite. The rock is almost wholly dark minerals, with some feldspar.

Some 1500 minerals can be identified by a knowledgeable person using only a small hand lens. This represents roughly half of the total number of mineral species. For the avid mineral collector, recognizing 1500 species is exciting, but it is not necessary to the general understanding of the story of the rocks – a few minerals can act as guideposts. The more you know, of course, the more interesting the story becomes, as subplots and counterplots are revealed.

It is not easy to decide what any rock should be called. There are many reasons for this naming difficulty, but basically the problem is simply that rocks do not yet have names that are generally agreed upon, as do plants, animals, and fossils. There are many ways of naming rocks – many "classifications" – and no one system is yet officially recognized, although international committees are now working on an official worldwide classification. To help you get at least some idea of what to call a rock, the key beginning on p. 203 has been constructed in a manner similar to those used by biologists. Bear in mind that this key is for the Sierra Nevada only; if you try to use it on rocks from other places, it may give you a wrong answer.

Even after you have arrived at a simple field name for a rock ("schist," for example), it does not mean you can read the geological literature with any facility. In the past century and a half that California rocks have been studied, the same type of rock – indeed, the same outcrop – has been given a host of different names by different investigators. This changing of names represents an increased knowledge of the rock itself, as well as an increase in knowledge of the whole science of rocks. But it does make reading about them extraordinarily difficult. To help you understand what you read, there is a list of rock names used in the Sierra Nevada beginning on p. 194. These names are referred to the more simplified field names used in this book. You will discover that "calc-hornfels" is the same rock you may already know as "limestone."

In addition, charts telling you where to see good examples of the various types of rocks are included in the following chapters. If you become familiar with the rocks in the places suggested, you will have a good start toward recognizing the rocks of the Sierra Nevada.

Table 1

Common Rock-Forming Minerals

The few minerals that are essential to the field identification of igneous rocks are given here. The table is undoubtedly too simple, since it is not easy to be certain that a mineral is correctly identified when only a handful of the 1500 recognizable by sight are listed. A good mineral identification book is an excellent start toward correct identification, but the quickest, surest, and best way to learn minerals is to see and touch the minerals themselves.

Family	How to identify	Where to see an example
Quartz The most durable family of common minerals	May sparkle like tiny glass beads (vitreous luster) Knife or nail will not scratch Will not effervesce	Bear River, near Colfax (map 4, circle 5); cobbles of white quartz glisten in stream bed White quartz vein near Jamestown (map 8, circle 9) and Coulterville (map 8, circle 12)

Family	How to identify	Where to see an example
	("fizz") with acid Commonly is clear, especially if it is a constituent of granitic rocks but pebbles and boulders of white (milky) quartz are common in the Sierra Nevada	Clear crystals of quartz in mines near Mokelumne Hill (map 7, circle 10) Smoky crystals at Dinkey Dome (map 11, circle 25)
Feldspar Orthoclase group contains potassium; plagioclase group contains sodium and calcium The most common family of minerals	Nearly as hard as quartz; does not appear as glassy Resembles porcelain Orthoclase may be pink; plagioclase is white to dark gray Will not fizz in acid Many fragments of feldspar show a multitude of closely spaced parallel lines	Lembert Dome (map 9, circle 13) and Cathedral Peak (map 10, circle 4), Yosemite National Park; the "lumps" in the granitic rock are feldspar crystals Kingsbury grade, at the sharpest curve (map 3, circle 4) (see listing for mica) Twin Lakes, Fresno County (map 11, circle 19)
Mica Two varieties are common: muscovite, a clear mica, and biotite, which is dark An unusual mica in the Sierra Nevada is mariposite, a green mica containing chrome	All mica splits into very thin sheets Very soft and flaky Will not fizz in acid Glimmers in the sun, so that tiny fragments in stream bottoms may be mistaken for gold (which is much heavier)	Large "books" of mica are in pegmatite rock on Kingsbury grade, at the sharpest curve (east of Lake Tahoe)(map 3, circle 4) Muscovite is "isinglass;" screw-in fuses have muscovite plates in the ends Mariposite is in road outcrops on State Highway 49 near Coulterville (map 8, circle 13), and at the Josephine (map 8, circle 18) and Mary Harrison (map 8, circle 14) mines

Family	How to identify	Where to see an example
Amphibole Common in dark-colored igneous rocks	Commonly dark green or black; may look as if it were lacquered Crystals commonly long and narrow In granitic rock, commonly shows as brilliant black laths Breaks (cleaves) at oblique angles (56° and 124°); pyroxene cleaves at right angles	Carson Hill mine, Calaveras County (map 7, circle 21) Cosumnes copper mine, near Fairplay (map 5A, circle 3) Near Twin Lakes, Fresno County (map 11, circle 19) West of Vallecito (map 8, circle 4) West of Nevada City (map 4, circle 1)
Pyroxene Common in dark-colored igneous rocks	Resembles amphibole except that pyroxene breaks at nearly right angles, and crystals are short and stubby, rather than long and thin. An end view of a pyroxene crystal is nearly a square	In gabbro, Nevada City (map 4, circle 1) Twin Lakes, Fresno County (map 11, circle 19)
Calcite	Pearly or glassy appearance Breaks (cleaves) at oblique angles into rhombs Will effervesce ("fizz") if acid is placed on it Can be scratched by knife or nail	Common in metamorphic rocks in the Sierra Nevada Not common in Sierran igneous or sedimentary rocks See *calcareous rock*, table 2

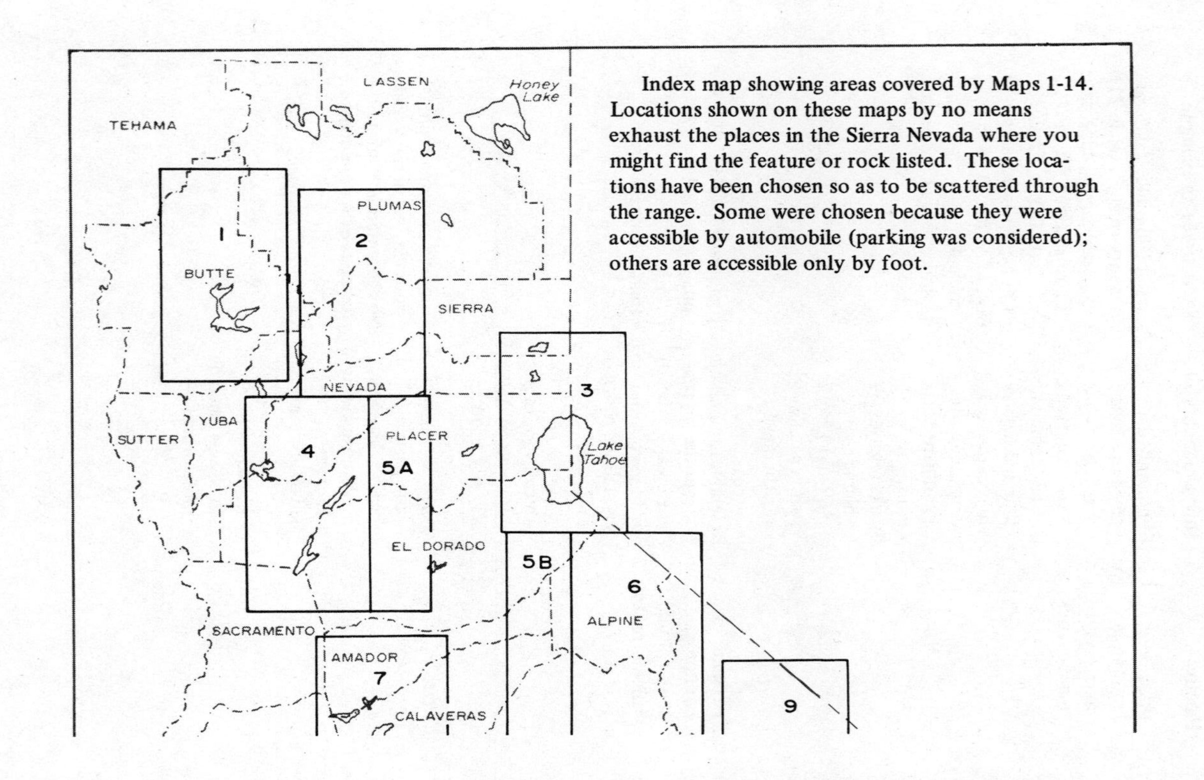

Index map showing areas covered by Maps 1-14. Locations shown on these maps by no means exhaust the places in the Sierra Nevada where you might find the feature or rock listed. These locations have been chosen so as to be scattered through the range. Some were chosen because they were accessible by automobile (parking was considered); others are accessible only by foot.

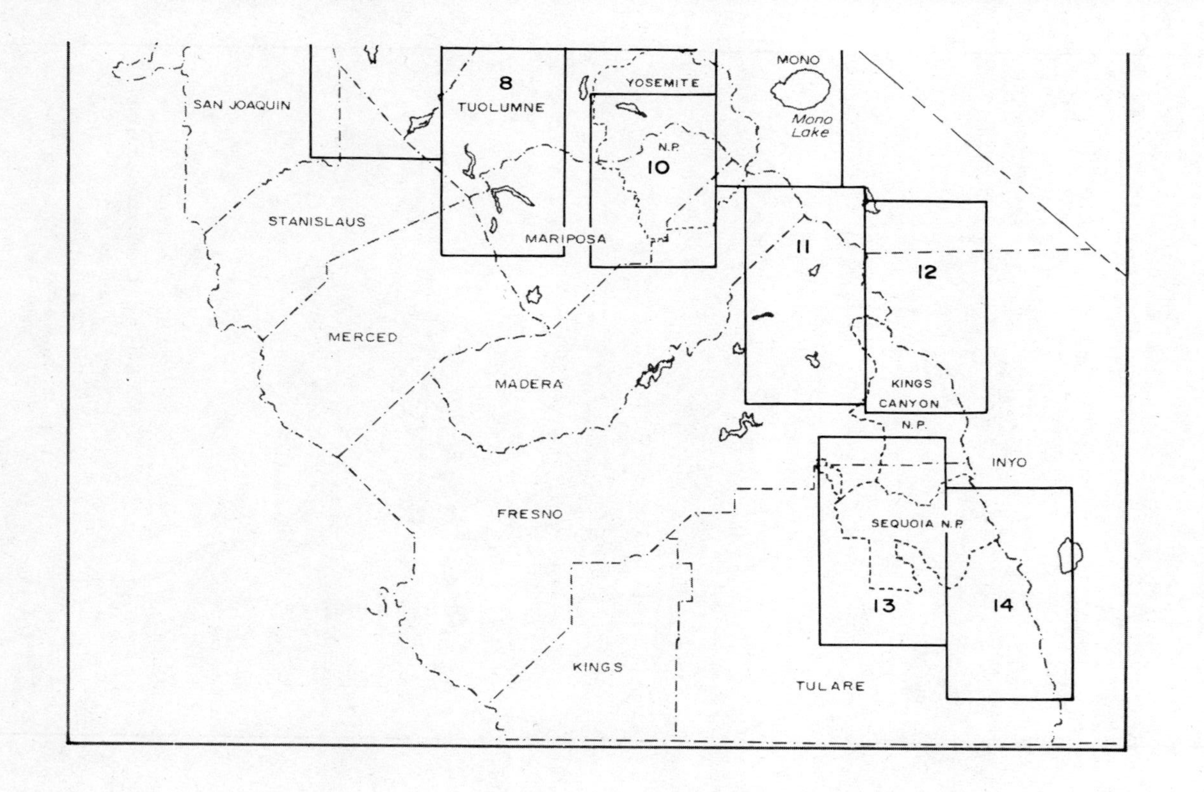

MONO
Mono Lake
8
TUOLUMNE
YOSEMITE
N.P.
10
SAN JOAQUIN
STANISLAUS
MARIPOSA
11
12
MERCED
MADERA
KINGS
CANYON
N.P.
INYO
SEQUOIA N.P.
FRESNO
13
14
KINGS
TULARE

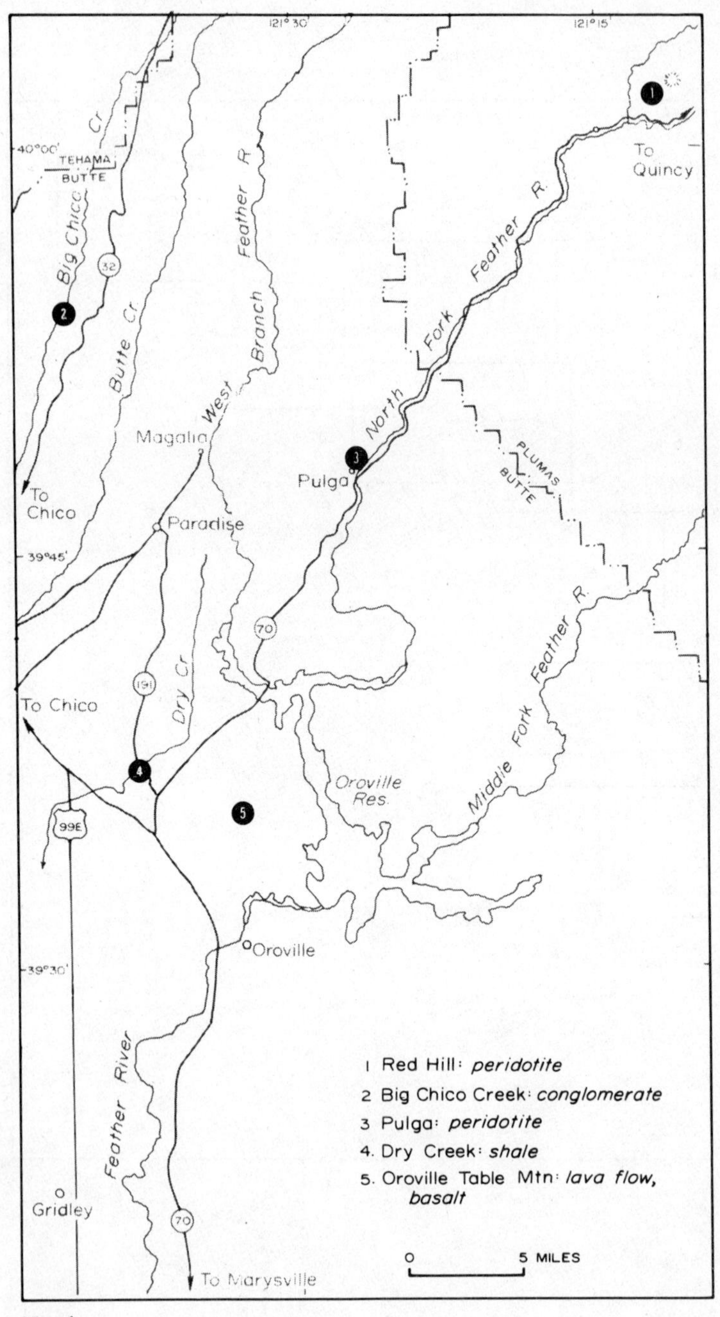
121°30'
121°15'
40°00'
TEHAMA
BUTTE
Big Chico Cr.
32
Butte Cr.
West Branch Feather R.
North Fork Feather R.
To Quincy
Magalia
Pulga
PLUMAS
BUTTE
To Chico
Paradise
39°45'
70
191
Dry Cr.
To Chico
99E
Oroville Res.
Middle Fork Feather R.
Oroville
39°30'
Feather River
Gridley
70
To Marysville
1 Red Hill: *peridotite*
2 Big Chico Creek: *conglomerate*
3 Pulga: *peridotite*
4. Dry Creek: *shale*
5. Oroville Table Mtn: *lava flow, basalt*
0 5 MILES

Map 1.

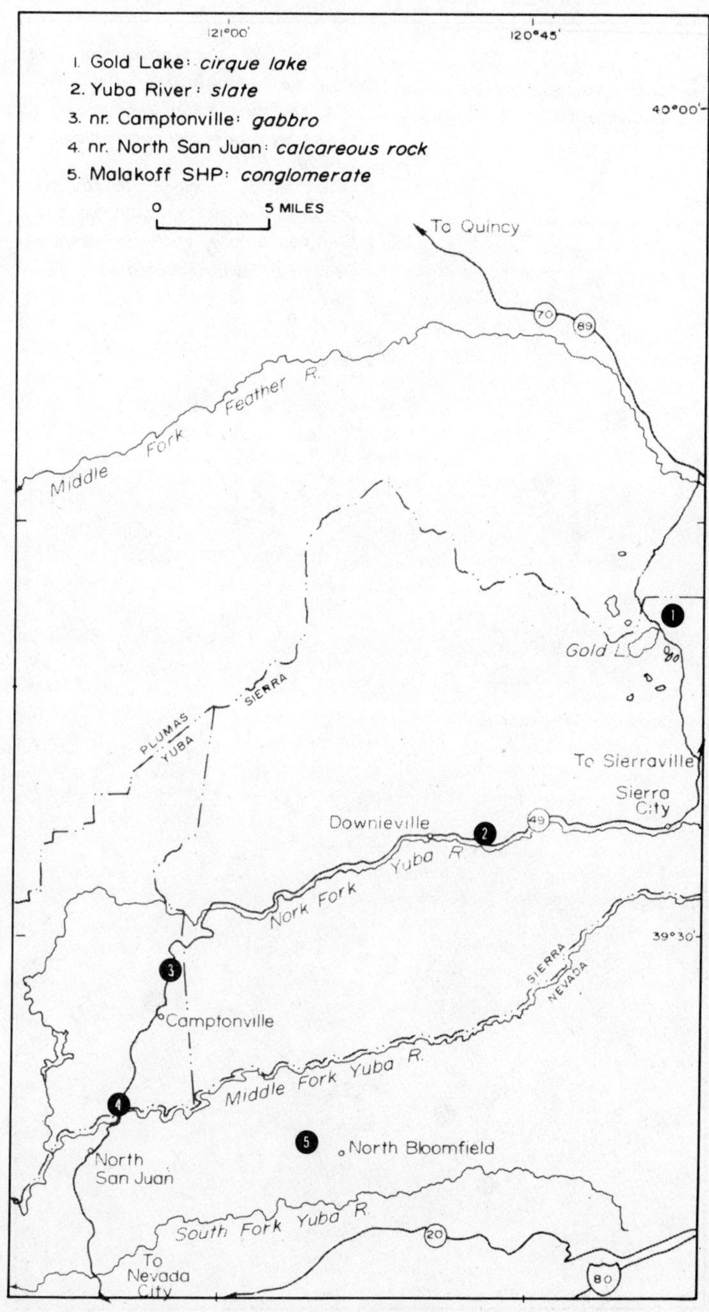

121°00'
120°45'
40°00'
39°30'
1. Gold Lake: cirque lake
2. Yuba River: slate
3. nr. Camptonville: gabbro
4. nr. North San Juan: calcareous rock
5. Malakoff SHP: conglomerate
0
5 MILES
To Quincy
70
89
Middle Fork Feather R.
Gold L.
PLUMAS
YUBA
SIERRA
To Sierraville
Sierra City
Downieville
49
North Fork Yuba R.
SIERRA
NEVADA
Camptonville
Middle Fork Yuba R.
North Bloomfield
North San Juan
South Fork Yuba R.
20
80
To Nevada City

Map 2.

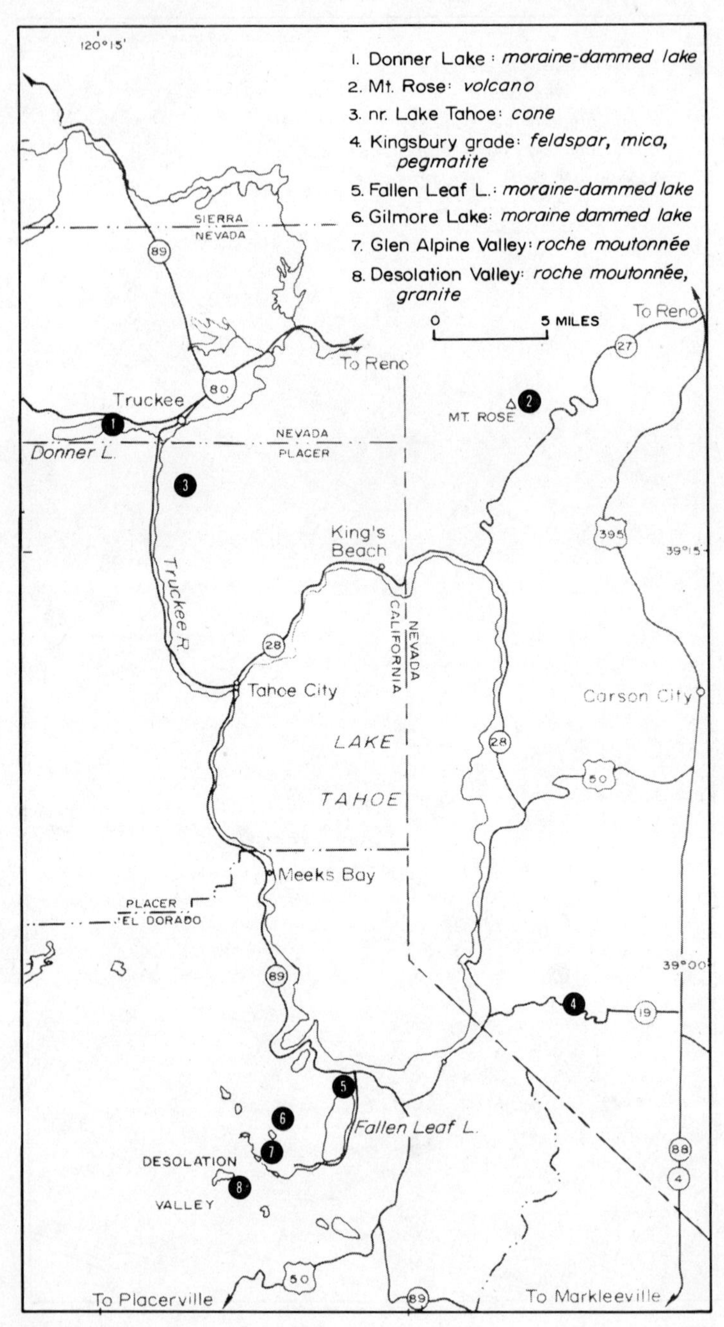
120°15'
1. Donner Lake: moraine-dammed lake
2. Mt. Rose: volcano
3. nr. Lake Tahoe: cone
4. Kingsbury grade: feldspar, mica, pegmatite
5. Fallen Leaf L.: moraine-dammed lake
6. Gilmore Lake: moraine dammed lake
7. Glen Alpine Valley: roche moutonnée
8. Desolation Valley: roche moutonnée, granite
0
5 MILES
SIERRA
NEVADA
89
To Reno
To Reno
27
80
Truckee
MT. ROSE
Donner L.
NEVADA
PLACER
King's
Beach
395
39°15'
Truckee R.
NEVADA
CALIFORNIA
28
Tahoe City
Carson City
LAKE
28
50
TAHOE
Meeks Bay
PLACER
EL DORADO
39°00'
89
19
Fallen Leaf L.
88
DESOLATION
4
VALLEY
50
To Placerville
89
To Markleeville
Map 3.

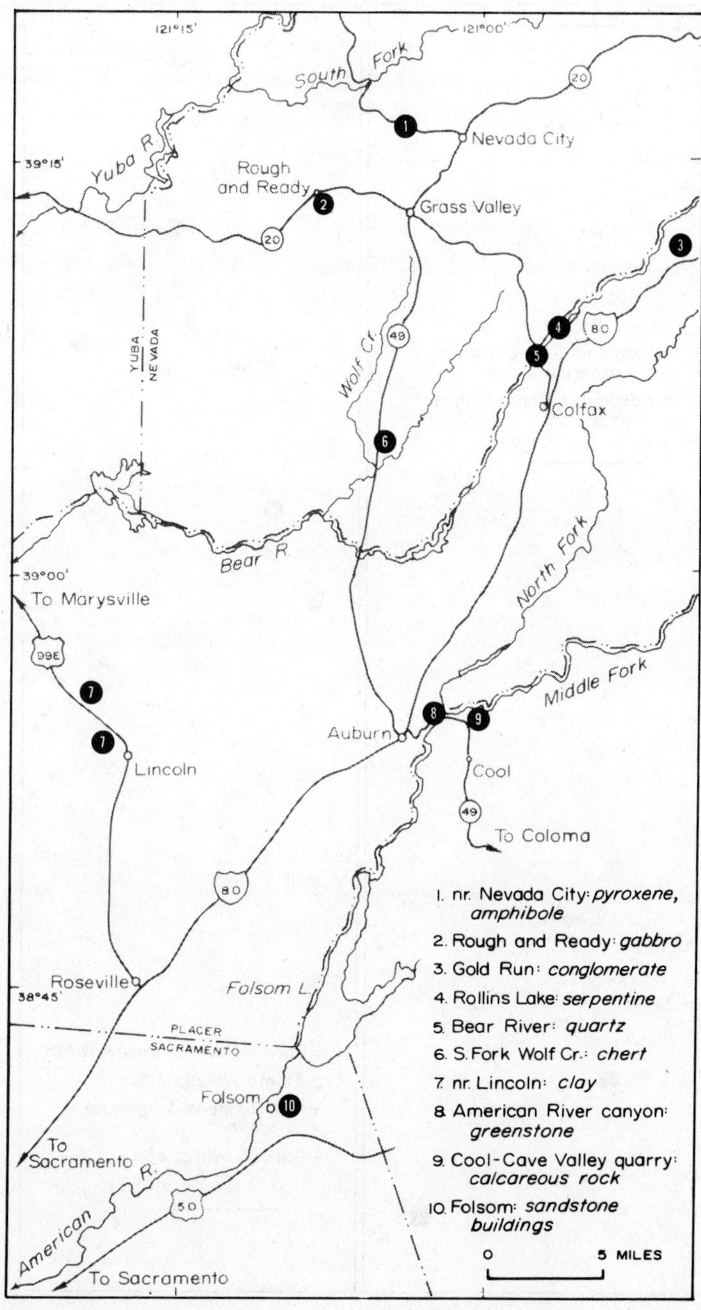
121°15'
121°00'
South Fork
Yuba R.
39°15'
Nevada City
Rough
and Ready
Grass Valley
20
49
80
Wolf Cr.
YUBA
NEVADA
Colfax
Bear R.
39°00'
To Marysville
North Fork
99E
Middle Fork
Auburn
Lincoln
Cool
To Coloma
Roseville
38°45'
Folsom L.
PLACER
SACRAMENTO
Folsom
To
Sacramento
American R.
50
To Sacramento
1. nr. Nevada City: pyroxene, amphibole
2. Rough and Ready: gabbro
3. Gold Run: conglomerate
4. Rollins Lake: serpentine
5. Bear River: quartz
6. S. Fork Wolf Cr.: chert
7. nr. Lincoln: clay
8. American River canyon: greenstone
9. Cool-Cave Valley quarry: calcareous rock
10. Folsom: sandstone buildings
0
5 MILES

Map 4.

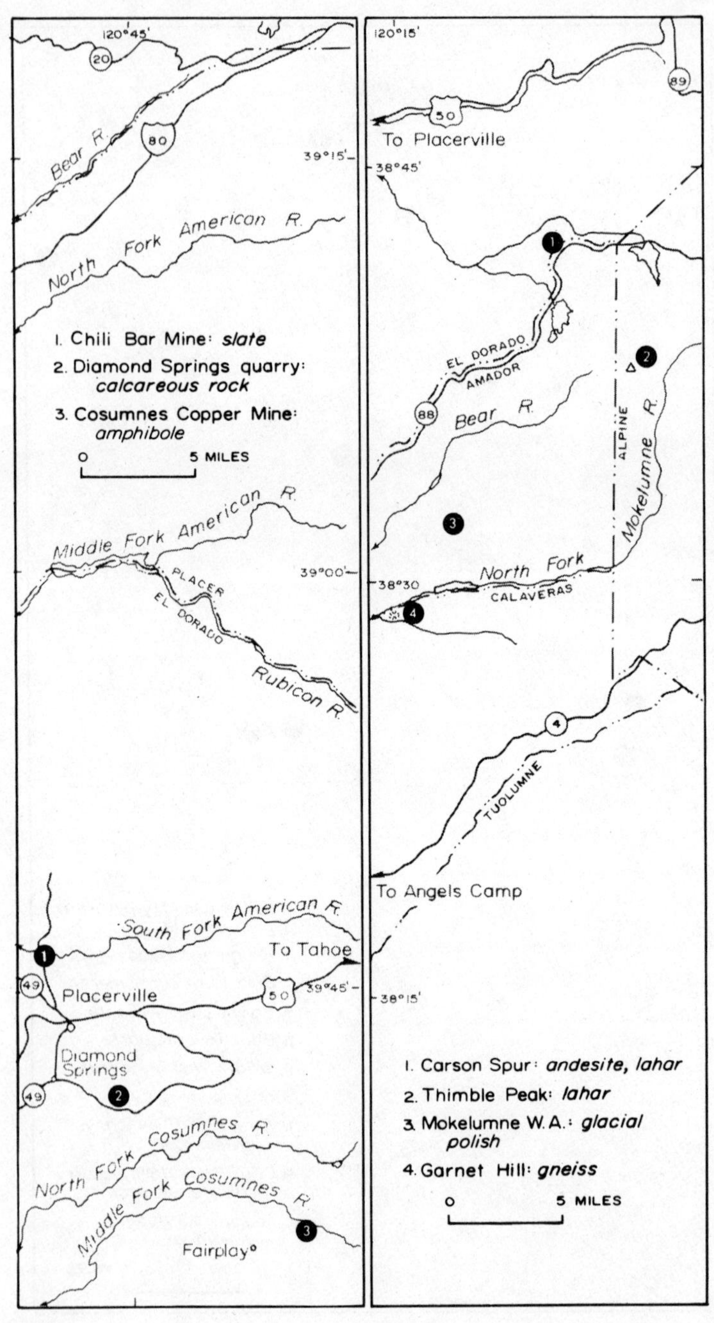
120°45'
20
80
Bear R.
39°15'
North Fork American R.
1. Chili Bar Mine: slate
2. Diamond Springs quarry: calcareous rock
3. Cosumnes Copper Mine: amphibole
0
5 MILES
Middle Fork American R.
39°00'
PLACER
EL DORADO
Rubicon R.
South Fork American R.
To Tahoe
49
Placerville
50
39°45'
Diamond Springs
North Fork Cosumnes R.
Middle Fork Cosumnes R.
Fairplay
120°15'
89
50
To Placerville
38°45'
EL DORADO
AMADOR
88
Bear R.
ALPINE
Mokelumne R.
North Fork
38°30
CALAVERAS
4
TUOLUMNE
To Angels Camp
38°15'
1. Carson Spur: andesite, lahar
2. Thimble Peak: lahar
3. Mokelumne W.A.: glacial polish
4. Garnet Hill: gneiss
0
5 MILES

Map 5A.

Map 5B.

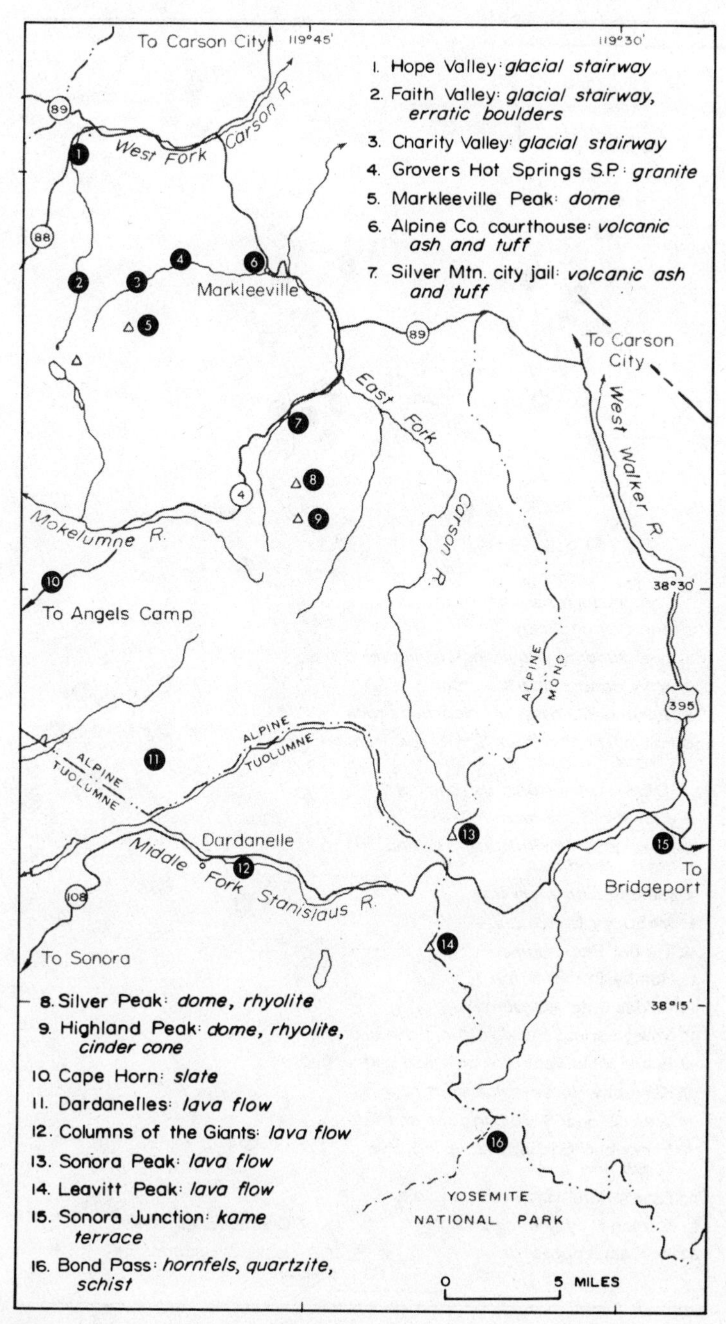

To Carson City
119°45'
119°30'
1. Hope Valley: glacial stairway
2. Faith Valley: glacial stairway, erratic boulders
3. Charity Valley: glacial stairway
4. Grovers Hot Springs S.P.: granite
5. Markleeville Peak: dome
6. Alpine Co. courthouse: volcanic ash and tuff
7. Silver Mtn. city jail: volcanic ash and tuff
89
West Fork Carson R.
88
Markleeville
To Carson City
East Fork
West Walker R.
4
Mokelumne R.
Carson R.
38°30'
To Angels Camp
ALPINE
MONO
395
ALPINE
TUOLUMNE
Dardanelle
Middle Fork Stanislaus R.
To Bridgeport
108
To Sonora
38°15'
8. Silver Peak: dome, rhyolite
9. Highland Peak: dome, rhyolite, cinder cone
10. Cape Horn: slate
11. Dardanelles: lava flow
12. Columns of the Giants: lava flow
13. Sonora Peak: lava flow
14. Leavitt Peak: lava flow
15. Sonora Junction: kame terrace
16. Bond Pass: hornfels, quartzite, schist
YOSEMITE NATIONAL PARK
0 5 MILES

Map 6.

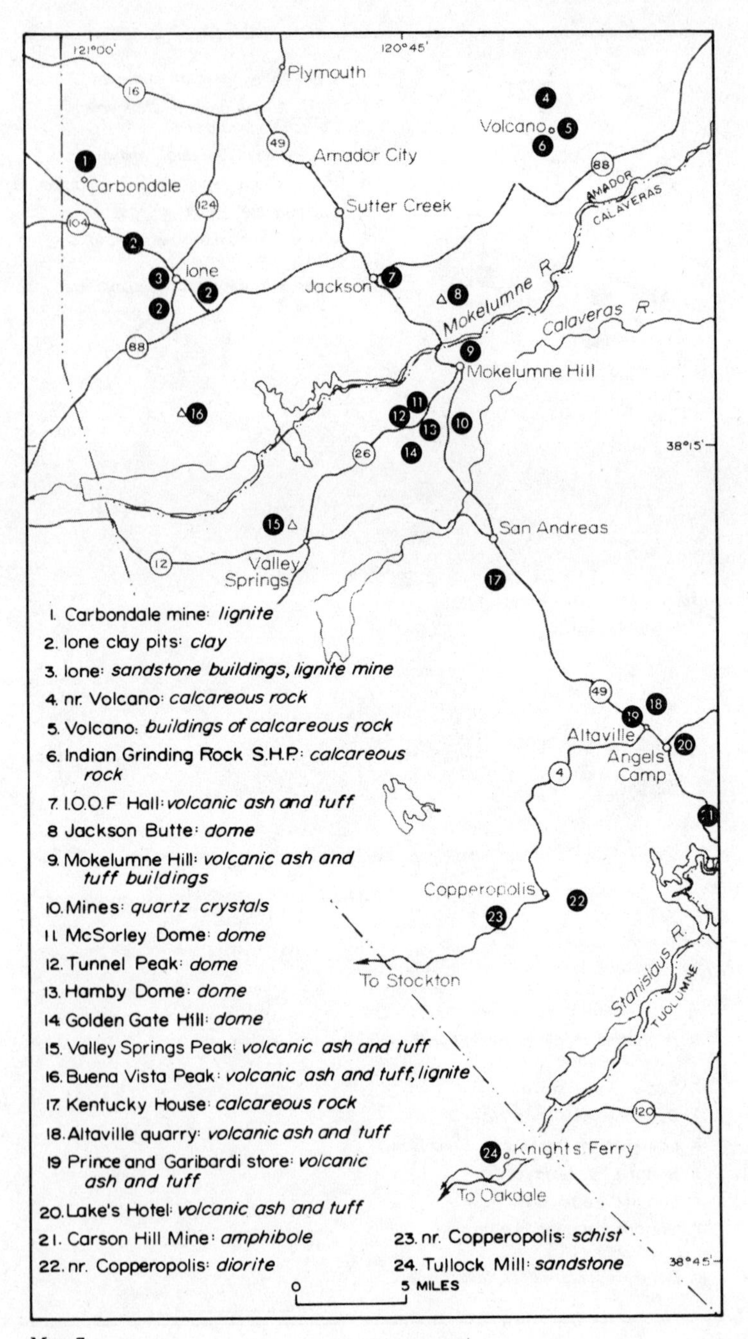
121°00'
120°45'
Plymouth
16
49
Amador City
Volcano
88
Carbondale
124
Sutter Creek
AMADOR
CALAVERAS
104
Ione
Jackson
Mokelumne R.
Calaveras R.
88
Mokelumne Hill
26
38°15'
San Andreas
12
Valley
Springs
49
Altaville
Angels
Camp
4
Copperopolis
To Stockton
Stanislaus R.
TUOLUMNE
120
Knights Ferry
To Oakdale
38°45'
0
5 MILES
1. Carbondale mine: lignite
2. Ione clay pits: clay
3. Ione: sandstone buildings, lignite mine
4. nr. Volcano: calcareous rock
5. Volcano: buildings of calcareous rock
6. Indian Grinding Rock S.H.P.: calcareous rock
7. I.O.O.F Hall: volcanic ash and tuff
8 Jackson Butte: dome
9. Mokelumne Hill: volcanic ash and tuff buildings
10. Mines: quartz crystals
11. McSorley Dome: dome
12. Tunnel Peak: dome
13. Hamby Dome: dome
14 Golden Gate Hill: dome
15. Valley Springs Peak: volcanic ash and tuff
16. Buena Vista Peak: volcanic ash and tuff, lignite
17. Kentucky House: calcareous rock
18. Altaville quarry: volcanic ash and tuff
19 Prince and Garibardi store: volcanic ash and tuff
20. Lake's Hotel: volcanic ash and tuff
21. Carson Hill Mine: amphibole
22. nr. Copperopolis: diorite
23. nr. Copperopolis: schist
24. Tullock Mill: sandstone

Map 7.

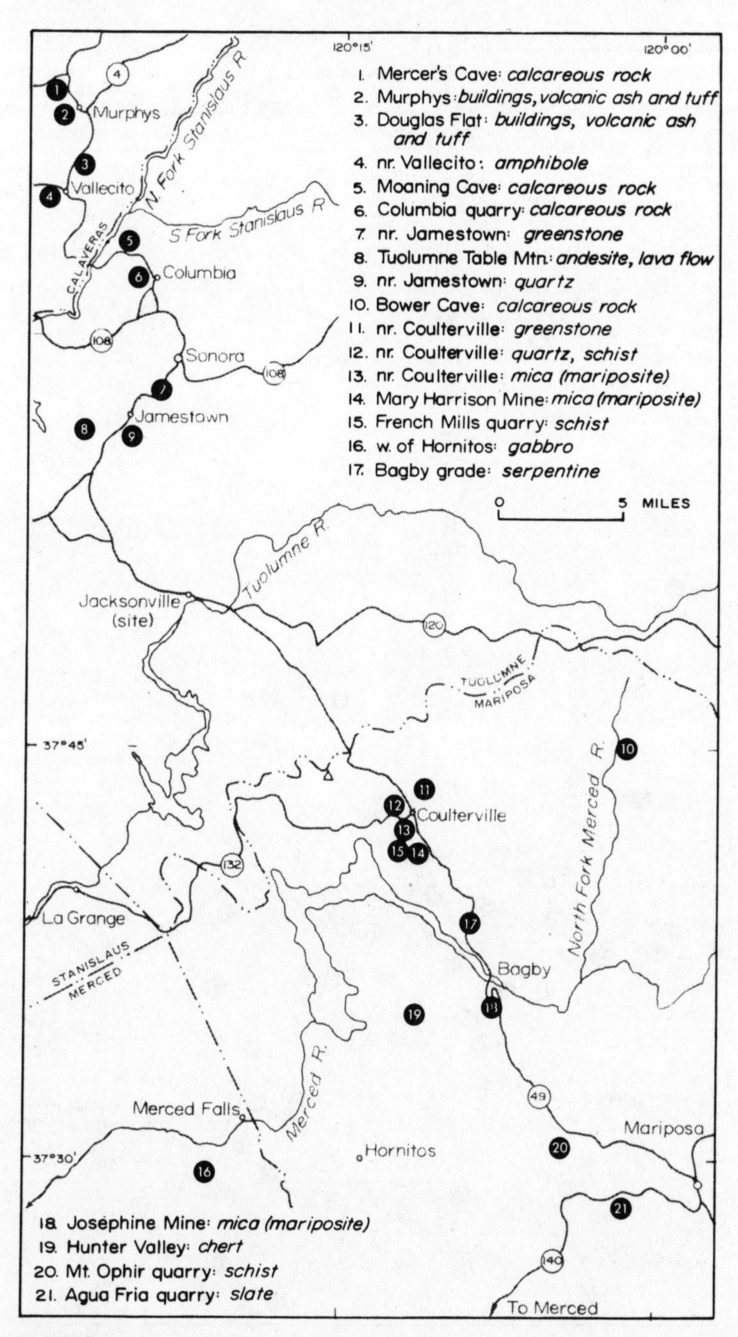
120°15'
120°00'
1. Mercer's Cave: calcareous rock
2. Murphys: buildings, volcanic ash and tuff
3. Douglas Flat: buildings, volcanic ash and tuff
4. nr. Vallecito: amphibole
5. Moaning Cave: calcareous rock
6. Columbia quarry: calcareous rock
7. nr. Jamestown: greenstone
8. Tuolumne Table Mtn.: andesite, lava flow
9. nr. Jamestown: quartz
10. Bower Cave: calcareous rock
11. nr. Coulterville: greenstone
12. nr. Coulterville: quartz, schist
13. nr. Coulterville: mica (mariposite)
14. Mary Harrison Mine: mica (mariposite)
15. French Mills quarry: schist
16. w. of Hornitos: gabbro
17. Bagby grade: serpentine
0 5 MILES
Murphys
Vallecito
N. Fork Stanislaus R.
S. Fork Stanislaus R.
CALAVERAS
Columbia
Sonora
Jamestown
Tuolumne R.
Jacksonville (site)
TUOLUMNE
MARIPOSA
37°45'
Coulterville
North Fork Merced R.
La Grange
STANISLAUS
MERCED
Bagby
Merced R.
Merced Falls
Hornitos
Mariposa
37°30'
To Merced
18. Josephine Mine: mica (mariposite)
19. Hunter Valley: chert
20. Mt. Ophir quarry: schist
21. Agua Fria quarry: slate

Map 8.

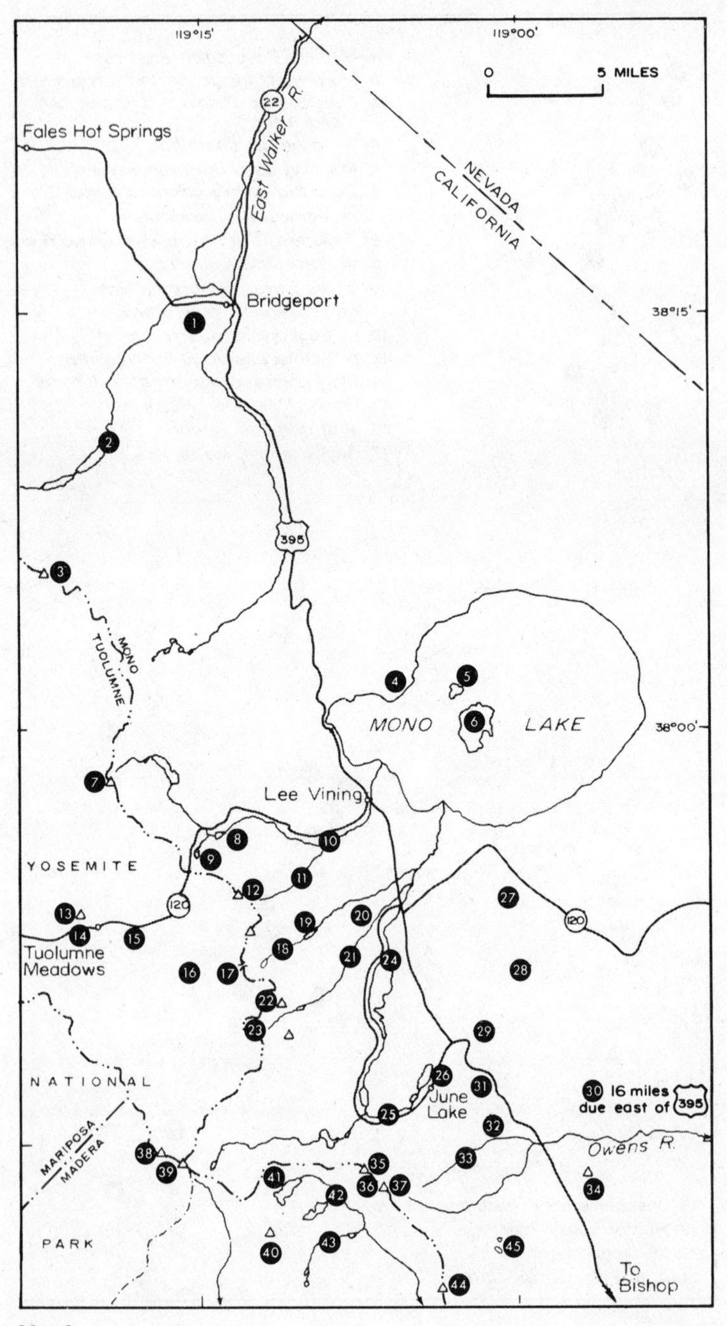

119°15'
119°00'
0
5 MILES
Fales Hot Springs
22
East Walker R.
NEVADA
CALIFORNIA
Bridgeport
38°15'
395
MONO
TUOLUMNE
MONO
LAKE
38°00'
Lee Vining
YOSEMITE
120
Tuolumne
Meadows
120
June
Lake
16 miles
due east of
395
Owens R.
NATIONAL
MARIPOSA
MADERA
PARK
To
Bishop

Map 9.

1. nr. Bridgeport: *till*
2. Twin Lakes: *moraine*
3. Matterhorn Peak: *matterhorn*
4. Black Pt.: *cone*
5. Negit Island: *volcano*
6. Paoha Island: *cone, explosion pits*
7. Mt. Conness: *bergschrund*
8. Ellery Lake: *hornfels*
9. Tioga Lake: *hornfels*
10. Lee Vining Canyon: *moraine*
11. Gibbs Canyon: *moraine*
12. Mt. Dana: *bergschrund, terminal moraine*
13. Lembert Dome: *feldspar, porphyry, glacial polish*
14. Tuolumne Meadows: *glacial moulin work, roche moutonnée*
15. Sand Meadows: *outwash plain*
16. Kuna Crest: *cirque lake*
17. Mono Pass: *hornfels, col*
18. Bloody Canyon: *till, glacial polish, scratches, grooves*
19. Walker Lake: *moraine, moraine-dammed lake, till*
20. Sawmill Canyon: *moraine, till*
21. Parker Creek: *perched boulders*
22. Mt. Lewis: *hornfels*
23. Kuna glacier: *terminal moraine*
24. Grant Lake: *moraine-dammed lake, chatter mark, scratches, grooves*
25. Reversed Creek: *till*
26. June Lake: *erratic boulders, moraine-dammed lake*
27. Panum Crater: *dome*
28. Mono Craters: *dome, obsidian, pumice*
29. Devil's Punchbowl: *pumice*
30. Glass Mountain: *dome*
31. Wilson Butte: *rhyolite, dome*
32. Obsidian Dome: *obsidian*
33. Glass Creek: *lava flow*
34. Lookout Mtn.: *rhyolite*
35. San Joaquin Mtn: *lava flow*
36. Cirque between 35 and 37: *rock glacier*
37. Two Teats: *lahar*
38. Mt. McClure: *bergschrund, arête*
39. Mt. Lyell: *bergschrund, arête, terminal moraine*
40. Ritter Range: *arête*
41. Thousand Island Lake: *cirque lake*
42. Garnet Lake: *cirque lake*
43. Shadow Canyon: *schist*
44. Minarets Lookout: *hornfels*
45. Inyo Craters: *explosion pits*

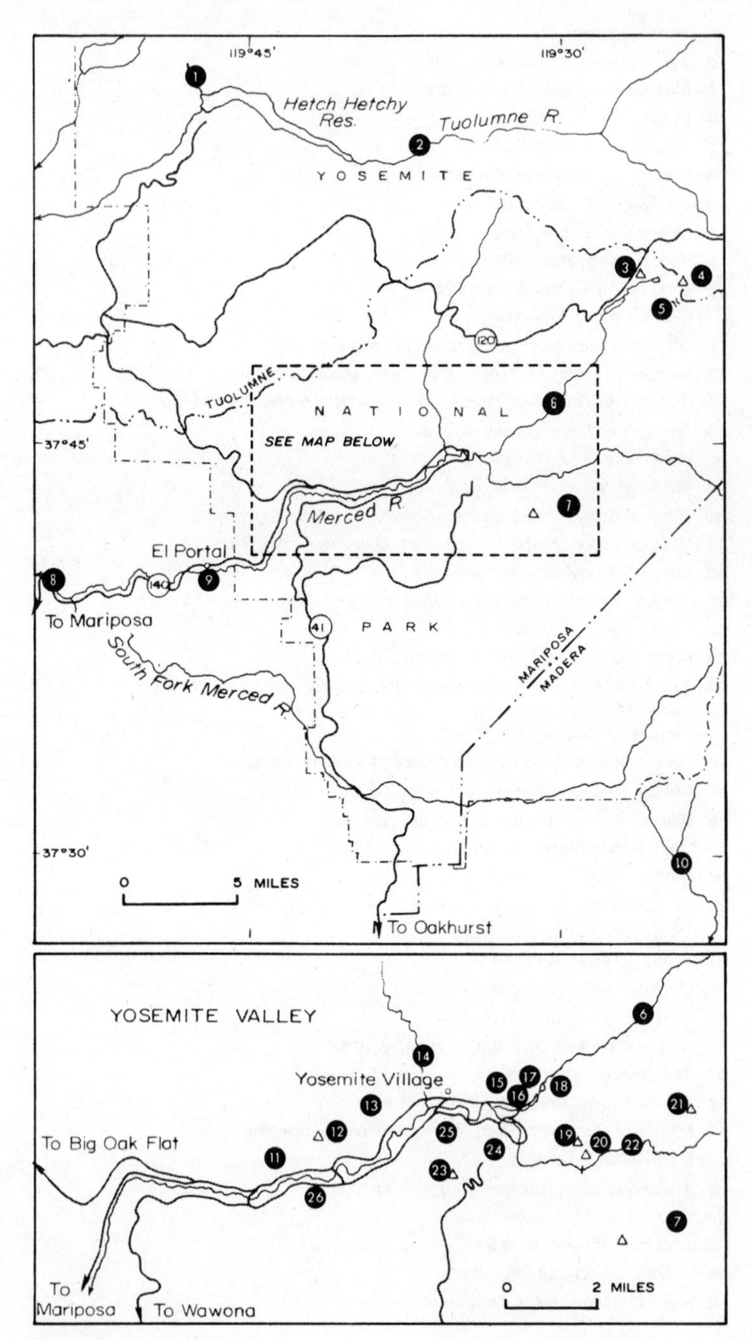
119°45'
119°30'
Hetch Hetchy Res.
Tuolumne R.
YOSEMITE
120
TUOLUMNE
NATIONAL
SEE MAP BELOW
37°45'
Merced R.
El Portal
140
To Mariposa
41
PARK
South Fork Merced R.
MARIPOSA
MADERA
37°30'
0
5 MILES
To Oakhurst
YOSEMITE VALLEY
Yosemite Village
To Big Oak Flat
To Mariposa
To Wawona
0
2 MILES

Map 10.

1. Tueeulala Falls: *hanging valley*
2. Hetch Hetchy Valley: *U-shaped valley*
3. Pywiack Dome: *glacial polish*
4. Cathedral Peak: *feldspar, porphyry*
5. Cathedral Pass: *scratches, grooves*
6. Tenaya Canyon: *glacial polish*
7. Starr King Meadows: *perched boulders, erratic boulders*
8. Geological Exhibit: *phyllite, chert*
9. El Portal: *diorite*
10. Chiquito Creek: *erratic boulders, perched boulders*
11. The Rockslides: *diorite*
12. El Capitan: *glacial polish*
13. Three Brothers: *glacial polish*
14. Upper Yosemite Falls: *perched boulders*
15. Royal Arches: *scratches, grooves*
16. Indian Grinding Rock: *granite*
17. Washington Column: *glacial polish*
18. Mirror Lake: *glacial polish*
19. Mt. Broderick: *glacial polish, roche moutonnée*
20. Liberty Cap: *glacial polish, roche moutonnée*
21. Moraine Dome: *erratic boulders, perched boulders*
22. Upper Merced Canyon: *glacial polish*
23. Sentinel Dome: *erratic boulders*
24. Glacier Point: *erratic boulders*
25. Union Point: *glacial polish*
26. Cathedral Rocks: *erratic boulders*

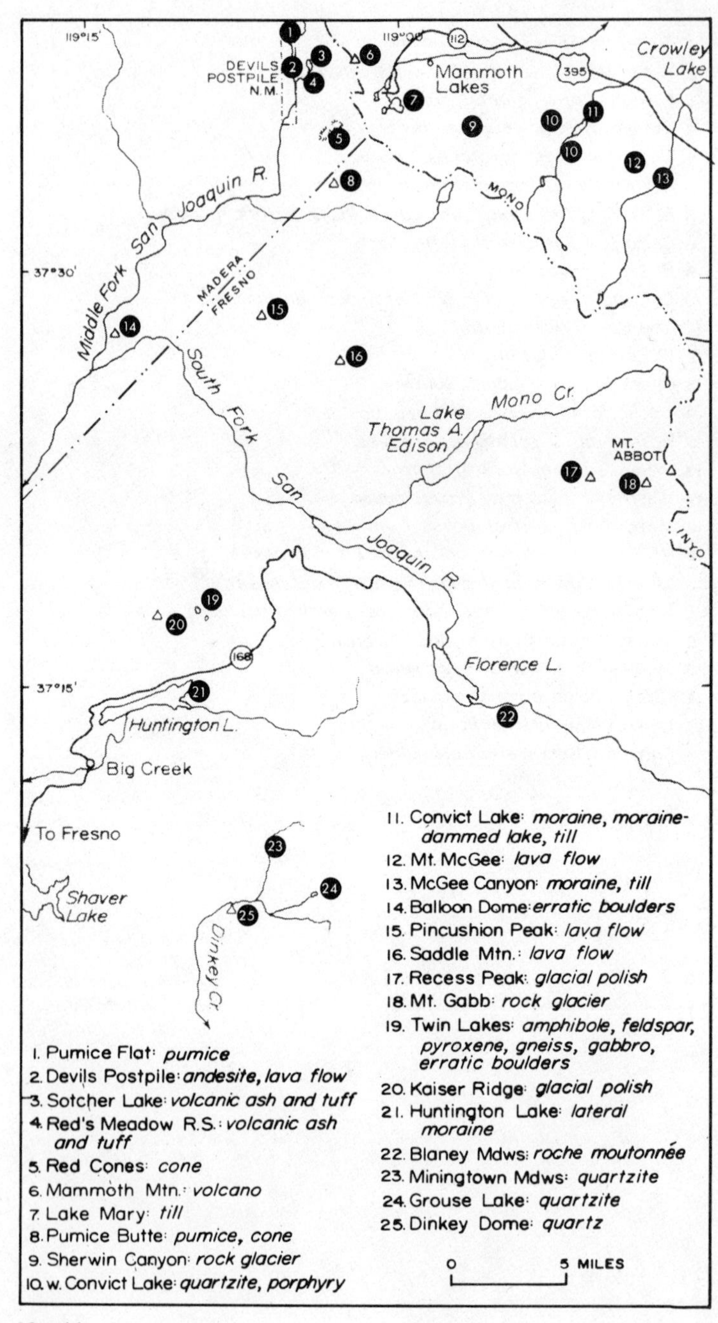
119°15'
119°00'
112
395
Crowley Lake
DEVILS POSTPILE N.M.
Mammoth Lakes
San Joaquin R.
Middle Fork
MONO
37°30'
MADERA
FRESNO
South Fork
Lake Thomas A. Edison
Mono Cr.
MT. ABBOT
San
Joaquin R.
INYO
168
Florence L.
37°15'
Huntington L.
Big Creek
To Fresno
Shaver Lake
Dinkey Cr.
1. Pumice Flat: pumice
2. Devils Postpile: andesite, lava flow
3. Sotcher Lake: volcanic ash and tuff
4. Red's Meadow R.S.: volcanic ash and tuff
5. Red Cones: cone
6. Mammoth Mtn.: volcano
7. Lake Mary: till
8. Pumice Butte: pumice, cone
9. Sherwin Canyon: rock glacier
10. w. Convict Lake: quartzite, porphyry
11. Convict Lake: moraine, moraine-dammed lake, till
12. Mt. McGee: lava flow
13. McGee Canyon: moraine, till
14. Balloon Dome: erratic boulders
15. Pincushion Peak: lava flow
16. Saddle Mtn.: lava flow
17. Recess Peak: glacial polish
18. Mt. Gabb: rock glacier
19. Twin Lakes: amphibole, feldspar, pyroxene, gneiss, gabbro, erratic boulders
20. Kaiser Ridge: glacial polish
21. Huntington Lake: lateral moraine
22. Blaney Mdws: roche moutonnée
23. Miningtown Mdws: quartzite
24. Grouse Lake: quartzite
25. Dinkey Dome: quartz
0 5 MILES

Map 11.

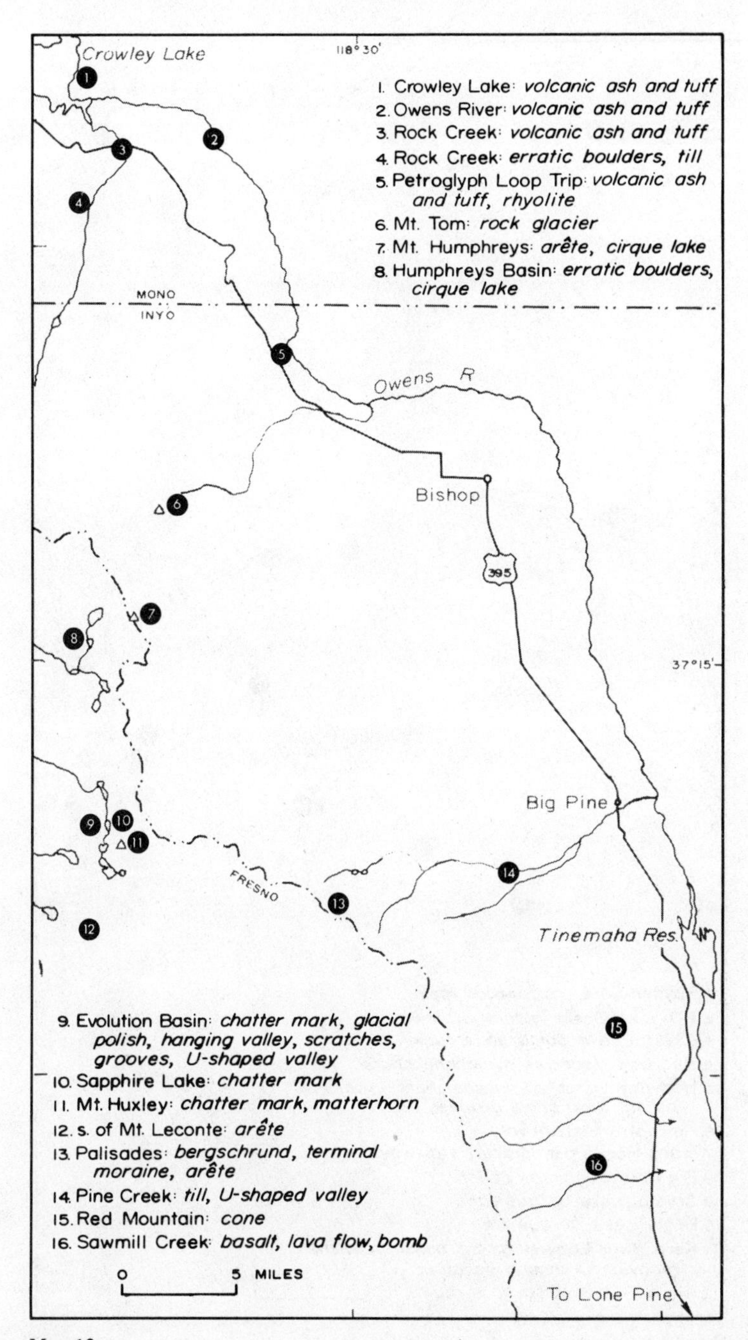

Crowley Lake
118° 30'
1. Crowley Lake: volcanic ash and tuff
2. Owens River: volcanic ash and tuff
3. Rock Creek: volcanic ash and tuff
4. Rock Creek: erratic boulders, till
5. Petroglyph Loop Trip: volcanic ash and tuff, rhyolite
6. Mt. Tom: rock glacier
7. Mt. Humphreys: arête, cirque lake
8. Humphreys Basin: erratic boulders, cirque lake
MONO
INYO
Owens R
Bishop
395
37°15'
Big Pine
FRESNO
Tinemaha Res.
9. Evolution Basin: chatter mark, glacial polish, hanging valley, scratches, grooves, U-shaped valley
10. Sapphire Lake: chatter mark
11. Mt. Huxley: chatter mark, matterhorn
12. s. of Mt. Leconte: arête
13. Palisades: bergschrund, terminal moraine, arête
14. Pine Creek: till, U-shaped valley
15. Red Mountain: cone
16. Sawmill Creek: basalt, lava flow, bomb
0 5 MILES
To Lone Pine

Map 12.

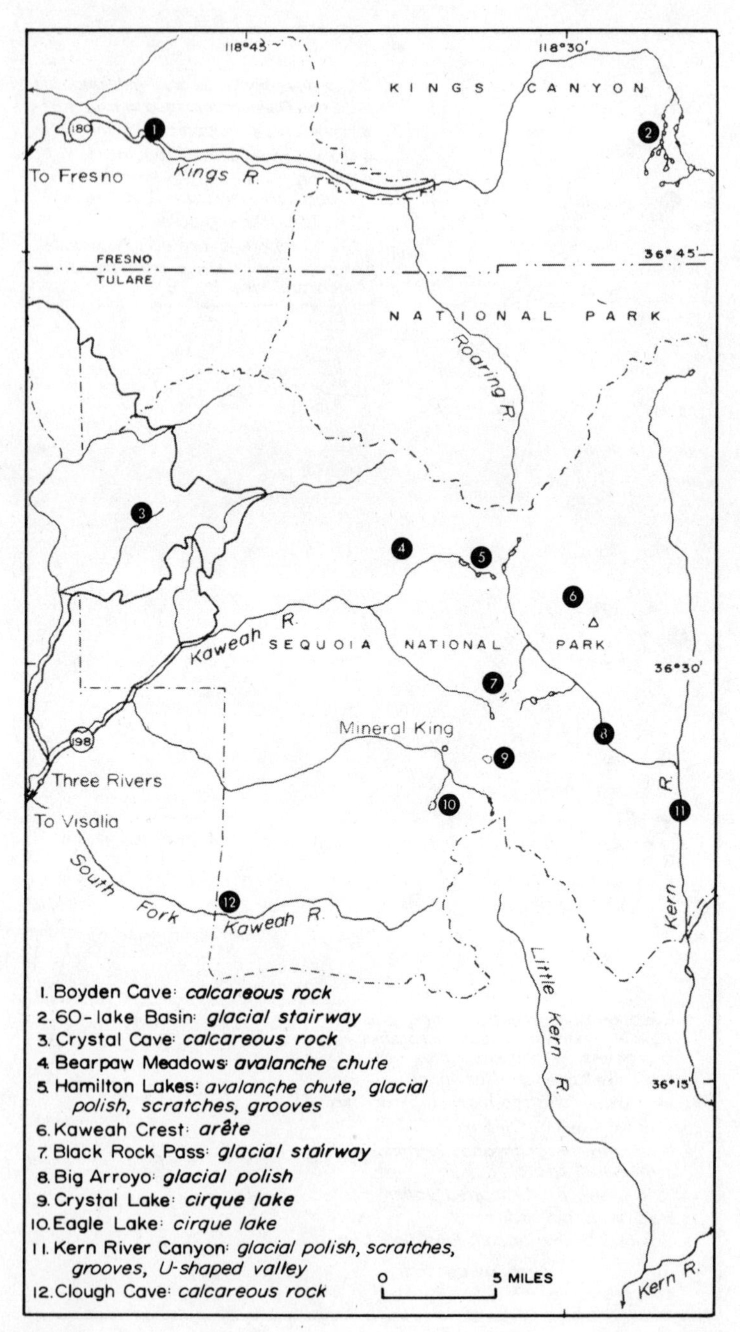

118°45'
118°30'
KINGS CANYON
180
To Fresno
Kings R.
FRESNO
TULARE
36°45'
NATIONAL PARK
Roaring R.
Kaweah R.
SEQUOIA NATIONAL PARK
36°30'
Mineral King
198
Three Rivers
To Visalia
South Fork Kaweah R.
Kern R.
Little Kern R.
36°15'
1. Boyden Cave: calcareous rock
2. 60-lake Basin: glacial stairway
3. Crystal Cave: calcareous rock
4. Bearpaw Meadows: avalanche chute
5. Hamilton Lakes: avalanche chute, glacial polish, scratches, grooves
6. Kaweah Crest: arête
7. Black Rock Pass: glacial stairway
8. Big Arroyo: glacial polish
9. Crystal Lake: cirque lake
10. Eagle Lake: cirque lake
11. Kern River Canyon: glacial polish, scratches, grooves, U-shaped valley
12. Clough Cave: calcareous rock
0 5 MILES

Map 13.

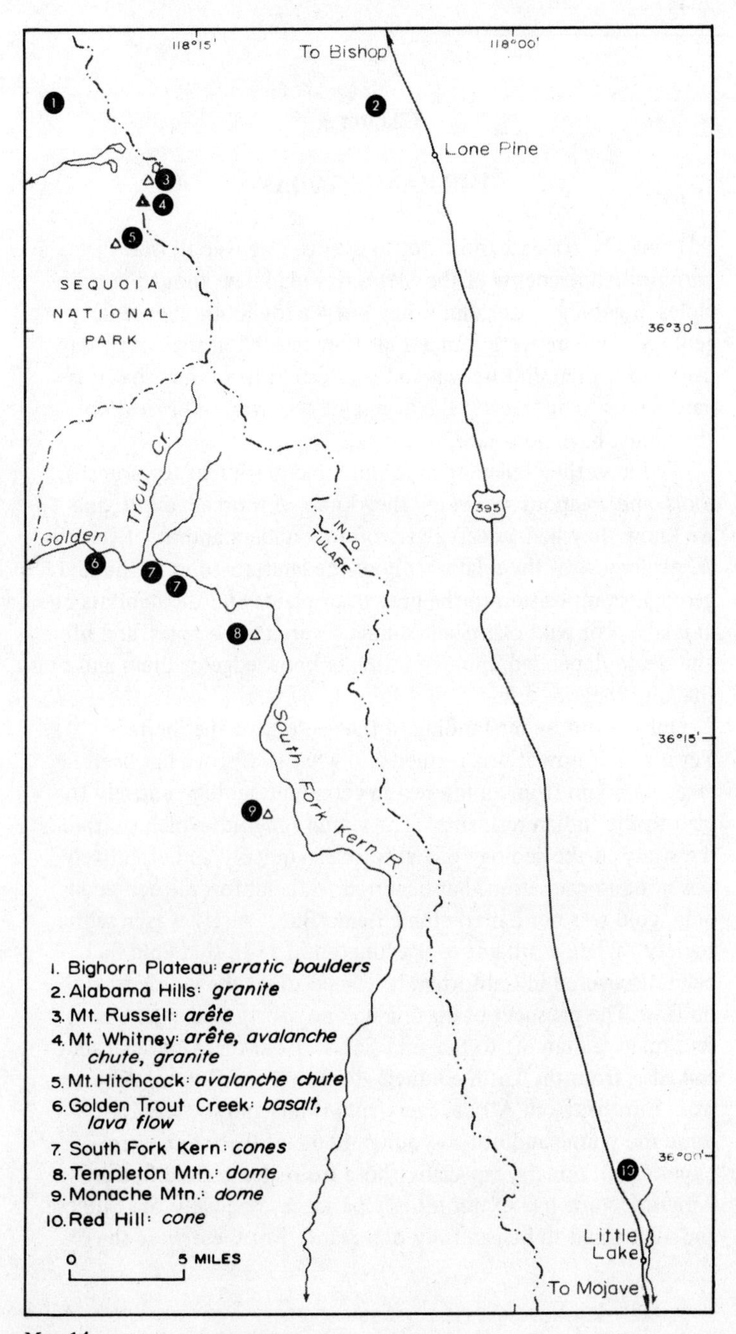

118°15'
To Bishop
118°00'
Lone Pine
SEQUOIA
NATIONAL
PARK
36°30'
Trout Cr.
Golden
395
INYO
TULARE
South Fork Kern R.
36°15'
36°00'
Little
Lake
To Mojave
1. Bighorn Plateau: erratic boulders
2. Alabama Hills: granite
3. Mt. Russell: arête
4. Mt. Whitney: arête, avalanche chute, granite
5. Mt. Hitchcock: avalanche chute
6. Golden Trout Creek: basalt, lava flow
7. South Fork Kern: cones
8. Templeton Mtn.: dome
9. Monache Mtn.: dome
10. Red Hill: cone
0
5 MILES

Map 14.

Chapter 3

THE RANGE TODAY

It may be too late, now, for us ever to discover all that the early Indian residents of the Sierra Nevada knew about the principles of geology. Certainly they knew a lot about economic geology – where to find materials they needed in their everyday lives – but their culture was so far different from ours that it is hard for us to appreciate the degree of geologic sophistication they must have possessed.

We know they knew of rock and mineral sources for utensils, tools, and weapons; we know they knew of sources of salt; and we know they had a deep and profound understanding of ecological geology – of the relationship of the land, its topography and geologic composition to the growth of plants and the habitats of animals. For wild plants and animals were their staples, and life and death depended upon an intimate knowledge of them and of the land they lived in.

Our present understanding of the geology of the Sierra Nevada – of how it was formed and what its history has been – takes its origin from an interest in economic geology entirely foreign to the Indian residents. The search for gold, which spurred the study of the geology of California so quickly and effectively, was not an occupation that occurred to them; for, in their economy, gold was not a marketable item. But it surely was in white society. When word got to Washington in 1848 that gold had been discovered in California, it seemed to set the whole world on fire. The prospect of digging up one's fortune in a new land sent many a man off to the gold fields: from the American south and east, from the Latin countries of South and Central America, from Europe, from Africa, even from China and the Far East came the young and not-so-young. Some of these men were experienced miners, especially those from Mexico and South America; some had extraordinary patience, especially the Chinese; and all learned to be carefully observant. For these were the qualities necessary to a good miner: experience, patience, and careful observation, mixed with a generous helping of good luck.

At first it seemed that Lady Luck alone panned with the miners. The first few months only rivers tumbling from the mountains were worked. Lucky prospectors stumbled upon huge nuggets and mined gravel containing $500 worth of gold in a single pan, while unlucky prospectors, working not far away, could recover only a few dollars, or a few cents, worth.

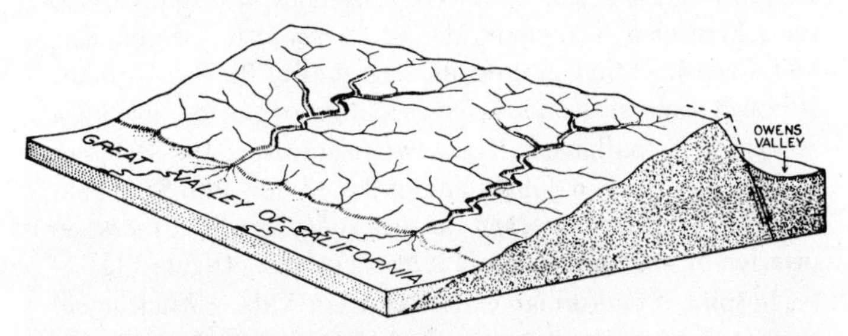

Fig. 10. Diagram of the tilted earth block that is the Sierra Nevada. The height and slant of the range are exaggerated. Streams are shown flowing in the general direction that Sierran streams flow, but no particular streams are intended. In front of the mountains is the Great Valley of California, comprised of the Sacramento and San Joaquin valleys, filled with sediment derived from the mountains. Owens Valley is marked on the east side of the mountain block. The fault system that bounds the east face is marked by arrows that indicate the direction of movement. (See Chapter 10.)

Gradually, the more canny and experienced among them began to discover some design to the accumulation of gold. It was obvious that the gold was in stream beds, and therefore that a knowledge of the habits of streams might be as useful to the prospector as the habits of deer were to the Indian. Later, they were to learn that a knowledge of the habits of today's streams would teach them about the habits of fossil streams, and lead them to the discovery of even greater accumulations of gold.

It is a political and geographic fact, dependent upon geology, that almost all California rivers rise in California and end in the sea off the California coast or sink into the ground. There are a few exceptions: the Colorado rises in the Rocky Mountains and flows into the Gulf of California; the Truckee, Carson, and Walker Rivers rise in the Sierra Nevada and flow into Nevada. The Owens River, whose bed lies at the foot of the Sierra on the east, is – or rather, was – another exception, as its many tributaries rose in the High Sierra and flowed east, then south into Owens Lake. Today, much of the water that once flowed in the Owens River is sent through aqueducts to Los Angeles.

In the Sierra, the Feather, Yuba, Bear, and American Rivers all flow into the Sacramento. The Cosumnes, Mokelumne, Calaveras, Stanislaus, Tuolumne, Merced, Fresno, San Joaquin, Kings, and Kaweah all flow into the drainage area of the San Joaquin, although some of them lose their identity long before making a recognizable confluence. These two major rivers, the Sacramento and the San Joaquin, join in Suisun Bay to flow into San Pablo Bay, San Francisco Bay, and the sea. Today, much of the water is diverted or impounded before it reaches the sea (figure 11).

In spite of California's current concern with a sufficiency of water, a great deal falls on the Sierra Nevada, mostly in the form of snow. Much of the rain and snow falls on the western side, because the mountain bulk, standing as high as 14,000 feet in some places, interrupts the clouds as they are blown across country by westerly winds. Most of the moisture in these clouds is removed as moist air is lifted up the western flank, so that little falls on the eastern side, making that face and the land beyond it a desert. There, the yearly rainfall averages from 5 to 15 inches, while on the western side there are many places that receive 50 inches of moisture a year. At Tamarack, in Alpine County, 450 inches of snow – more than 37 feet – is the average, and in the winter of 1906-7, 884 inches fell – almost 74 feet! Tamarack holds the record for the greatest snowfall in one calendar month in all of North America: 390 inches (32½ feet), in January 1911; it also holds the U.S. record for having the greatest depth of snow on the ground at any one time (not counting snowdrifts): 451 inches, nearly 37½ feet!

The eastern slope of the Sierra Nevada is precipitous, with lofty mountain peaks rising quickly and starkly. It is a large range, 50 to 80 miles wide, containing nearly as much area as the

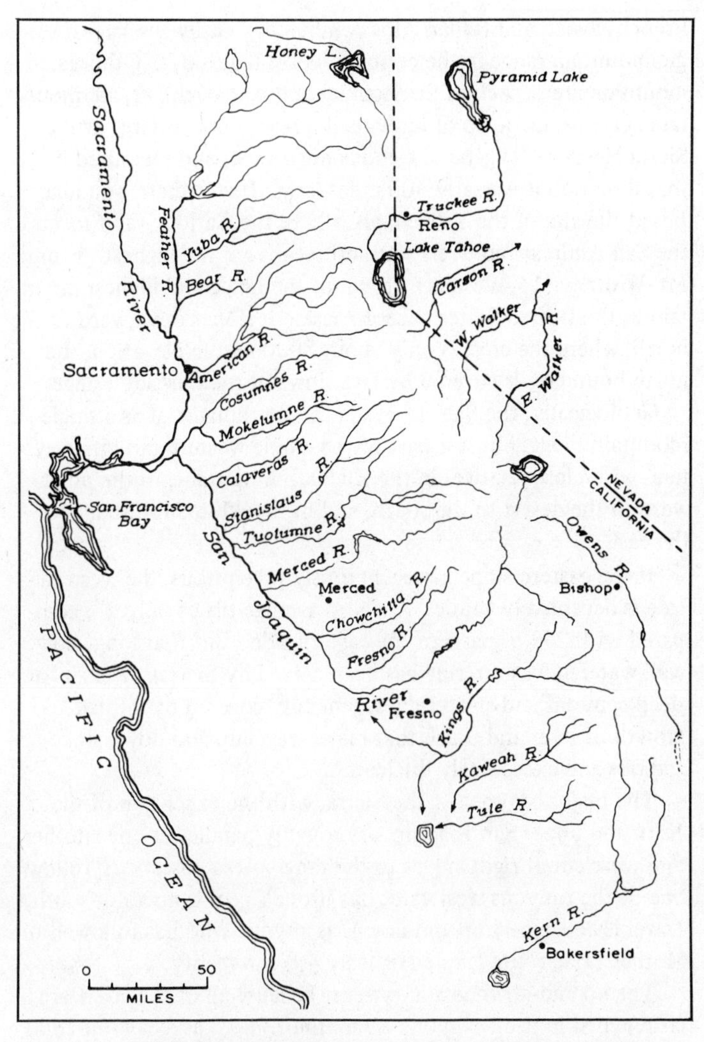

Fig. 11. Central California showing the major rivers that drain the Sierra Nevada. Most of them rise in the Sierra and flow into the San Joaquin or Sacramento and thence to the sea; a few flow east into Nevada.

French, Swiss, and Italian Alps combined – easily the largest single mountain range in the contiguous United States. Other large mountain areas, such as the Rockies and Appalachians, are mountain systems, made up of individual ranges, none as large as the Sierra Nevada. Its general trend is northwest, and measured in that direction it is nearly 400 miles long. Its southern boundary lies at the end of the Tehachapis, where the Garlock fault meets the San Andreas fault. Its southern peaks are the highest. From Mt. Whitney (14,496 feet), crown of the range and highest mountain in the United States except Alaska, it slopes downward to the north, where the crest is only about 8000 feet in elevation. Its north boundary is covered by lava flows of the Cascade Range.

Geologically, the Sierra Nevada can be thought of as a single mountain range, but it is part of the whole western mountain system, with close relatives in the Klamath Mountains to the northwest, in the desert to the south, and in the White Mountains to the east.

If the eastern slope is treacherously precipitous, the western side is deceptively gentle. It has an average tilt of only 2°, compared with 25° of parts of the eastern side. But that long, slow, well-watered western side has been carved by Sierran rivers into deep canyons and steep ridges, generally covered by a thick growth of trees and brush that make travel up and down slope tedious and exceedingly difficult.

The major canyons of the Sierra, with the exception of the Kern and upper San Joaquin, are roughly parallel to one another, and were cut at right angles to the range. It is possible to follow one of the canyons westward; but if one's goal is north or south, travel is constantly up and down, as anyone who has followed the Mother Lode route (State Highway 49) can testify.

The up-and-downness is extreme because all of the rivers are entrenched in deep canyons. One, the Kings, has a canyon 7000 feet deep – considerably deeper than the Grand Canyon. To the hiker, the numerous deep canyons and sharp ridges present chaos. It is not apparent to him that the entire mountain block is tipped slightly southwestward, and that master streams generally flow in that direction (figure 10). In fact, they do not always flow in the principal downslope direction. In the Mother Lode and along the northern mines, where the streams leave the granite of the high country to enter the metamorphic terrain of the foothills, they deviate from their courses, forced into a new direction by the resistant ridges of old rock. Water, like all else in the universe, conserves its own energy. It bends its course when it strikes obstacles, seeking the easiest, though not necessarily the shortest, route to the sea.

As the streams pass through the metamorphic rocks, they pick up particles of rock, including gold, that have broken loose in the fierce Sierran winter. Such was the gold that the early Argonauts sought – particles of gold carried along by the water, resting here and there in a stream channel on their way westward. The miner who could predict where the gold was likely to accumulate by studying the stream was more apt to be successful than one who depended wholly upon luck.

As anyone who has used a garden hose knows, the faster and heavier the stream of water, the greater the amount of soil and leaves it will push. Other things being equal – which they rarely are – this can be expressed as the "load" of a stream, the amount of rock, sand, and soil that it can move. Obviously, streams flow faster down steep slopes than they do down gentle ones. In general, then, one would expect streams in steep, high mountains to be stronger and straighter than those in more level country, as they have more strength to push aside obstacles in their paths than do the sluggish streams of the lower slopes. There are meanders – those repeated bends in stream channels – where streams are flowing over nearly level ground. The Mississippi in its lower reaches meanders widely; there are miniature Mississippi patterns in the Sierra in high mountain flats, such as Leavitt Meadows east of Sonora Pass.

Almost all rivers tend to form constantly winding paths – to begin to meander. Even if water is started down a clean sheet of glass, the path the water takes begins to curve, just as raindrops may wander down a vertical window. The less steep the slope the water flows down, the greater the number of bends. There are, in fact, meanders in the fresh-water currents that traverse the world's oceans.

The reason is not immediately apparent. True, the stream deflects its course as it strikes objects, but to the careful observer, it seems that the stream "overreacts" – that its bends are greater than one would expect from the size of the obstacle. Indeed, if one were to study the patterns of streams on topographic maps, he might conclude that the smaller the obstacle, the greater the deflection!

A river meanders because it is seeking the easiest route to follow. A stream that meanders uses less energy than one that does not. Therefore, where streams have little speed and force, the easiest course for them to follow is a meandering one – for them, the longest way 'round is the easiest way home!

It is this tendency for streams to bend that provided the resting points for the particles of gold that prospectors sought and still seek. The stream carries a great mass of sand and gravel – and the gold with it – in its bed. Where the stream bends, or where there are irregularities in the bottom of the bed, some of the load of rock particles is dropped, at least until water from a new storm gives renewed vigor to the stream.

In general, the larger and heavier the particle, the harder it is for a stream to move it. There are enormous boulders in the stream courses of the high mountains that are in the process of being moved. Large boulders are bounced along the bottom; sand and gravel are churned in the bed; and fine particles are carried in suspension or are dissolved in the water. The muddiness of rivers that meander over wide flat areas is witness to the load they carry. The particles in suspension or solution cannot always be seen, but they may tint the water a light chocolate or milky color. The clarity of mountain streams, in contrast, testifies to their load: in the colorless water, large pieces – sand, gravel, boulders – lie on the bottom, with only a few rock particles small enough or flat enough to be suspended.

Eventually a river, if it has enough water, enters the sea. Where it enters, much of the load it has been carrying – small particles, because by this time the river is too sluggish to carry large ones – is dropped to form a delta. A delta is wedge-shaped, if viewed from the top, like the Greek letter from which it takes its name.

California's best known delta is not at the seashore. It is inland, where the Sacramento and San Joaquin rivers meet. This great region, criss-crossed by a thousand miles of river-channel waterways that meander and join in an intricate braided pattern, is the point at which Sierran water slows in its race downhill, before it mingles in ocean water.

Although this region where the two rivers merge is called the "Delta," it is not a simple triangle of the form and shape of the Nile delta, to which the word "delta" was first applied. It is, instead, two deltas; one a distributary for the San Joaquin River, one for the Sacramento. Rivers that enter the sea have deltas that fan out broadly. In contrast, the delta system of the Sacramento-San Joaquin is truncated. After the rivers' loads are dropped at the "Delta," the combined rivers then race through the Carquinez Straits, bend into San Francisco Bay, and move out the Golden Gate to the sea.

The story of this delta has been complicated by faulting, by the drowning of the area with sea water when ice melted at the end of the Ice Age, and by choking with debris from hydraulic gold mines in the Sierra Nevada.

The life story of rivers from the first drop of rain to their plunge into the sea or their disappearance into the desert sand is of great interest to all of us today. We value the brooks for their beauty; the rivers for their wildness. We thirst for their waters for outselves, our lands, and our machines, and need the gravel they carry and the life they support.

For all these many reasons, the "Delta," from the river confluence to the Golden Gate, has been used to study how rivers and bays the world over behave. Engineers and scientists have made a scale model of the whole region to use as an exact tool. It is housed in Sausalito, Marin County, and is open to the public. One can watch the tides in speeded-up action, see the water rise and fall, and measure the effect changes in the land or water will have on the entire system. Those who made the model, and those who use it (it can be rented to use for testing purposes) are not interested in gold (although their equipment could be used to study placer gold deposits), but in how streams behave. They need to know how fast a reservoir will fill with sediment and how to prevent silt from clogging an irrigation channel.

Scientists have assumed that mechanical laws only applied to sedimentation; that sandstone, shale, conglomerate – all were formed according to the rules of physics alone. For example, there is a mathematical expression for the rate at which particles fall through water, taking into account their size and surface area. A different "law" gives the relationship of the shape of the particles to the rate at which they fall (in general, the flatter the particle, the slower it falls). Together, then, they should tell us how fast mud, silt, and sand will build up in watercourses, a matter of interest to engineers who manage our water supplies and to geologists, who need to know how fast natural processes take place.

But nature has surprises. An eight-year study of the Delta-Mendota Canal, a part of the California water project, showed that physical laws alone do not account for all the accumulation of sediment. The canal is a concrete-lined ditch for 95 of its 113-mile length, carrying Sierran water from the Delta area southward. Its water is unusually turbid, owing partly to organic material picked up from extensive peat beds the canal passes, and partly to abundant plankton living in the water. The organic material is very light in weight, easily carried long distances. Its presence prevents the rapid settlement of the fine mineral particles that engineers had predicted.

On the other hand, when the canal was emptied of its water, stretches of unexpected sedimentary layers on the concrete liner were laid bare. Although the layers were horizontally stratified, as one would expect from settled particles, they showed a color change from light gray at the top to black at the bottom. In the black sediments at the bottom were many empty shells of the Asiatic clam, *Corbicula fluminea,* which had been accidentally introduced into this country. On the gray surface of the upper sedimentary layer were many live clams. In these clam-rich areas, the amount of sediment deposited was unusually large.

Engineers, geologists, and biologists jointly studying the sediments and the sedimentary process came to the conclusion that the sediments were taken from the turbid canal water that had been pumped through the body of the clams. The layers were gradually turning to rock – shale, organically produced through excretion by clams! How many other of our geologic processes, heretofore considered to be principally or wholly mechanical, have an unsuspected biological aspect?

Not all rocks and minerals of the same size are the same weight. Some are as light as pumice, which floats on water, and some are nearly as heavy as gold. The heavy particles, particularly gold, are hard for the stream to move. They are so much heavier than the rest of the sand and gravel that the stream is constantly pushing the other material over the reluctant gold, thereby settling it and other heavy minerals deeper and deeper in the stream bed.

Even novice prospectors soon found that these heavy gold particles, in the process of being winnowed by the stream to the bottom of its bed, dropped on curves – more on the inside than on the outside. They found "pay streaks" – concentrations of gold and other heavy minerals that had been sorted by the stream into zones within the sand body. They found that where the river channel bottom was smooth, such as in areas of polished granite, gold was less likely to accumulate than where there were irregularities to trap the gold. Here, behind the bumps and in the holes, they found nests of placer gold.

In the lower reaches of Sierran streams, where the rivers leave the granite of the high country to enter an area made of metamorphic rock, the channel bottom became a natural gold-concentrating machine. The metamorphic rocks have been upended in many places, although their original position must surely have been as horizontal as the sand in lakes of today. As the particles of gold were pushed downstream along the bottom of the channel, they dropped behind projections of the upended rock, sometimes burrowing into the weathering bedrock itself.

Enterprising miners, knowing the habits of rivers, turned aside large waterways to mine the river bed. For one river-moving enterprise, which diverted the Feather River for more than a mile, miners built a concrete trough to contain the river while they stole its gold.

It is no longer necessary to move rivers for the sake of gold; skin divers of today mine Sierran streams every summer with fin and mask, pry bar and suction tube. No one has yet grown rich, but it is a form of mining very close to the original meaning of the word placer – "pleasure."

Where do the rivers get the sand, gravel, and silt they are carrying toward the sea? Quite obviously, it is worn from the mountains by wind and weather. Sometimes, one can watch the mountain being worn away: landslides, rockfalls, mud flows – all are tearing the mountains down, making smaller pieces that rivers and streams can move toward the sea. As the processes of erosion continue, whether at the speed of a rock fall or at nearly undetectable rates, the mountains are carved by wind, water, and ice into a mass of hills and valleys. The exact shape of the hills and valleys, the distribution and size of rivers that drain them, and, in turn, the plants that clothe them are determined by a number of factors. The kind of rocks, the height of the mountains, the amount of rainfall, the number of rivers, the strength of the wind, the frequency of earthquakes, the severity of the climate, and the actions of animals and people all exert some influence on what mountains look like.

Since erosion is constantly at work, are the mountains actually being worn away? The answer is, they are; however, they may not be getting lower, since mountain-building forces at work in the Sierra may be keeping pace or exceeding the rate of erosion.

Measurements of the San Joaquin River show that now, when dams have curtailed much of its flow and canals control it, the San Joaquin is carrying away the Sierra at the rate of an inch per thousand years. Compare this with the Eel River of northern California, which is removing the Klamath Mountains at the rate of 40 to 80 inches per thousand years! This is the fastest erosion in the nation – fifteen times as fast as the Mississippi is eroding its borderlands.

Chapter 4

SEAS OF LONG AGO

The first four billion years of the region that has become the Sierra Nevada is veiled, as the rock record of the earth's beginnings is missing in these mountains. The first pages of Sierran history that remain for us to read begin in rocks laid down about 500 million years ago (Ordovician Period). There are other, earlier pages in the structurally related White and Inyo mountains to the east, which probably reflect happenings in the Sierra; but in the Sierra Nevada proper, this is the oldest now known.

What these rocks and their younger relatives tell us is that the sea washed over the land of the Sierra while the long ages rolled – probably for more than 400 million years. Virtually all of the older rocks in the Sierra Nevada (called "Bedrock" by miners; "Subjacent Series" by earlier geologists) are marine in origin – that is, they were formed at the bottom of a sea, though not in its deepest part. Judging by the kind of rocks and the few fossils in them, the eastern part of what is now the mountains was closer to shore than the western part during Paleozoic time – from about 500 million to 225 million years ago. Somewhere east of today's mountains lay land. Gradually, the sea receded, and during the ensuing 100 million years both the eastern and western parts of what is now the Sierra were in shallower water.

Very little is left of the creatures that lived in those ancient seas. Fossils are scarce in the Sierra, largely because most remnants of life were destroyed when the layers of rock that entombed them were pushed upward, bent, twisted, and faulted during the several episodes that marked the creation of the range.

Suppose you were to travel backward in time, and downward into the depths of the sea of 250 million years ago; you would surely have an altogether different view of the Sierra than we have today (figure 12). Perhaps you would see a few small fishes swimming among the reefs and darting behind tall stems; or perhaps not, for no fish remains are preserved in the Sierra. What seems to be

Fig. 12. Floor of the sea as it may have looked in late Paleozoic time, 300 million years ago, in the region where the Sierra stands today. Crinoids, bryozoans, brachiopods, straight and coiled cephalopods, starfish, snails, corals, sponges, and trilobites crowd the underseascape.

a waving forest on the floor of the shallow sea is made up, not of plants, but of animals. There are sea lilies (crinoids) with stems and what appear to be leaves. Their tops, stirring gently in the warm water, look very much like flowers. If you look closely at them, you will see that their stems as well as the petal-like parts of their "flowers" are made up of hard discs. When they die, these stony pieces fall slowly to the sea floor to be preserved in rock for millions of years. Nearby are corals, too, that form thick-stemmed "trees." They are, in fact, animal cities – colonies that may be connected to the sea floor by the bodies of their own dead, with only their outer parts alive.

For all that they look like a garden, these creatures do not live in soil, or transform sunlight into food, as land plants do. There is little of either soil or sunlight in the depths of the sea. The ocean bottom is sandy, and even the rocks that lie tumbled about are covered with a rough crust. This, too, is an animal – a "moss animal" (bryozoan) – covering the rock so completely that you are not sure which are "rocks" made entirely of animals and which have cores of stone.

You cannot actually see the billions of tiny creatures floating in and on top of the water. If you could, you would see that some have minute clamlike shells; others have unique whorls and chambers; while still others resemble beautiful crystalline pinwheels. Some are the young of bryozoans, minute colonies on their way to becoming new aggregations of animals. These colonies are produced sexually, by mating within the stem of the older colony, there to ripen before spinning out to found a new bryozoan city.

Others, the tiny, elaborate crystalline ones, are radiolaria, floating through the water in clouds so thick you can see them, even though you cannot distinguish any single individual. These, and the tiny shelled foraminifers, are being lured by the crinoids into their waving tentacles. The crinoids wash the minute ones through their bodies by the millions, using the built-in sieves in their flowerlike heads to sift out the tiny animals – now their dinners – then shoot the water back into the sea.

Moving slowly along the sea floor here on the shelf of the continent are large coiled cephalopods. They resemble snails; but there are snails here, too, and clams, recognizable as relatives of those of our own time. You know some of the animals, attached to rocks and coral, as "lampshells" (brachiopods), but there are far more of them than you have ever seen before.

As you move westward and downward through this ancient sea, the scene darkens, for sunlight cannot penetrate deeply into the water. You feel fewer creatures about you as you move down the steep continental slope. Suddenly there is light and a great turbulence; a rush of warm water engulfs you. You are heading toward an undersea volcano, erupting incandescent lava on the sea floor. The sea is boiling so furiously that you cannot quite see the source.

You stop moving down the slope, and travel across it instead. As you do, a mass of mud, sand, and rock shoots past you at perhaps eighty feet a second. It is a "seaslide" – a turbidity current – speeding downward, a tumbling mass of sand and rocky debris.

The danger is great. It is a good time to return to today to see more safely what has been preserved for us of those seas of yesteryear.

While the seas covered the land, they piled up layer upon layer of sedimentary debris – sand, mud, skeletons and shells of animals, pieces of older rock and gravel. As new layers were heaped upon older ones, the weight of the new ones pressed down upon the older ones. Weight and pressure, together with chemical cement carried in the sea water, gradually consolidated the loose sediment into hard rock.

How many thousands of feet of stony layers were built up in the slowly sinking sea as it filled with debris, we cannot now tell. Early prospectors could see that these ridges of old rock on the western side of the mountains, below the granite of the high peaks, must be older than the range itself, for they were twisted, knotted, and gnarled; their beds, once horizontal, now were upended.

Geologists of a century ago recognized that, in the gold belt, there were two great groups of ancient rocks, one group that was Paleozoic in age, and one Mesozoic. We know that the Paleozoic rocks are more than 225 million years old, and that the Mesozoic ones were formed more than 75 million years ago. They constitute the remaining evidence of what the Sierran landscape was in those long-gone ages.

All the rocks in the Sierran Paleozoic and Mesozoic systems are now metamorphosed. In many places, it is possible to tell what they once were; when this is possible, geologists have called them by their original names – limestone, shale, sandstone – rather than by their more correct metamorphic names. This has, perhaps, promoted an understanding of the geologic history of the range, but has confused those who were trying to learn to recognize the rocks. If you read geologic reports on the Sierra Nevada – especially those published recently by the U.S. Geological Survey – you should bear in mind that all the Paleozoic and Mesozoic rocks except those younger than the granite are metamorphosed, even though they may be designated by names that indicate their origin rather than what they are today.

The Paleozoic beds once were chiefly limestone, siltstone, shale, and mudstone, with some volcanic rocks mixed in. Today, many of them can simply be called "hornfels," which means that they are even- and fine-grained metamorphic rock. A good field name for a rock that breaks in layers along mica minerals is "schist;" "slate," for one that breaks along flat planes but does not show shiny mica flakes; "marble," if the rock is calcareous and crystalline; or "quartzite," if it has been derived from sandstone and is crystalline. There have been many, many names applied to the metamorphic rocks of the Sierra Nevada. A list of them starts on p. 194. If you look up an unfamiliar rock name in this list, you can discover what general type of rock it is, and what word is used in this book for its general field name.

The study of the process of metamorphism – literally, "change of form" – is in its infancy. This is somewhat surprising, because many of our most valuable mineral deposits have been the result of metamorphism. In the language of geology, the metamorphic process is described as the application of heat, pressure, or both, on rocks, changing them from one form to another. Obviously, this definition does not include many changes of form that the

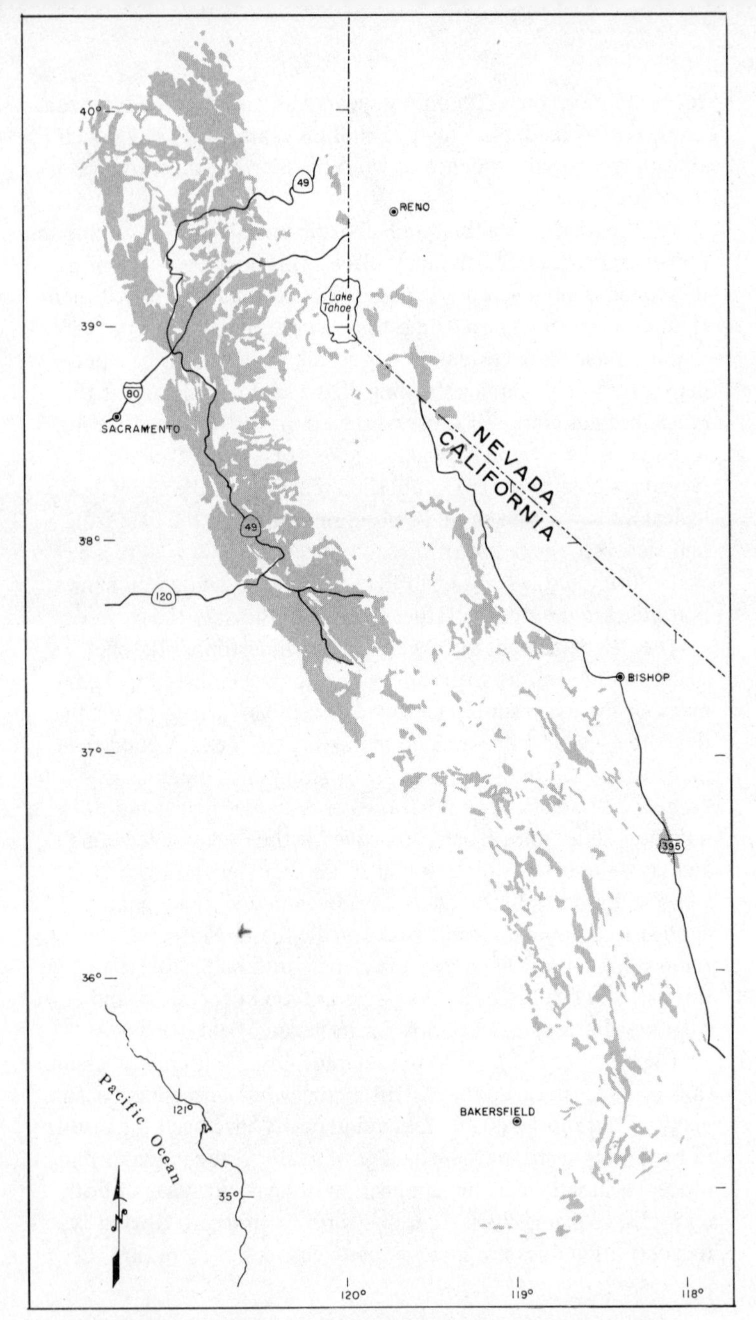

Fig. 13. Outcrops of Paleozoic and Mesozoic metamorphic rocks in the Sierra Nevada. These old rocks are now to be seen in the western foothills and along the crest of the range, although at one time they blanketed the entire area. Intrusion of granitic rock, uplift, and erosion have removed them from much of the range, allowing a view of the underlying granite.

rocks undergo, perhaps without the benefit of heat or pressure beyond that of the earth's surface. Weathering – the chemical change of minerals due to rain, snow, sun, wind, and other earth-surface forces – is surely a change of form. The rusting of metal is such a process. Yet these are all, largely for simplicity, not included as part of the metamorphic process. Neither is the process of consolidating sediments – mud, sand, gravel – into rock, although pressure and perhaps heat are surely involved. Nevertheless, in the broadest sense, all these processes are metamorphic.

At the other extreme, where heat or pressure is so great that some rocks within the depths of the earth are liquid and can digest other rocks completely, the metamorphic process is so complete that it has become igneous, involving fluid rock. We cannot then easily distinguish metamorphic rocks from igneous ones.

The "change of form" expressed by the word metamorphism is particularly a change in the chemical composition and form of the minerals that compose the rock itself. They change their form while solid, which gives them a different aspect from minerals crystallized from a liquid.

Commonly, metamorphic rocks tend to be "foliated" – to break along certain planes. This is due to the arrangement of metamorphic minerals which tend to flatten themselves against the pressure on them. In schist, for example, sheets of mica are arranged so that the easy way to flake them apart is parallel in all of the little mica books. This is in contrast to rocks that form from a liquid mass, where the pressure is equal in all directions. Mica books formed under these conditions may be oriented at random.

Mica is the most obviously oriented mineral, and the easiest to identify, but other minerals line up also. It is not that the minerals actually turn to become flattened against the pressure, but rather that the whole chemistry of the rock is reworked, using the old constituents (occasionally adding or subtracting some) to make wholly new minerals that grow in this flattened fashion.

In some places, where metamorphism has been intense, the rocks are wholly recrystallized, their original parents identifiable with great difficulty or not at all. Most Sierran rocks have not been this drastically changed.

In many Sierran rocks, the planes along which the rocks break are not straight, but wavy. These waves are clues to the rock's history. By painstaking geometric analysis, some geologists have undertaken to decipher the direction, and, to some extent, the amount of pressure the rock has undergone at various times. The few Sierran studies that have been made using this exacting technique indicate that the story of these old rocks is quite complex.

The older metamorphic rocks of Mesozoic age – those that formed before the granite backbone – are to be found in the same general areas as the still older Paleozoic beds (see figures 13 and 15). It is very difficult to tell the two groups apart unless they lie adjacent to one another. Where they do, a close look will show that the older, Paleozoic group is tilted at a different angle than the younger, Mesozoic strata. This lack of parallelism marks a time of unrest in the earth, when mountain-building forces were at work. It is technically called an "unconformity."

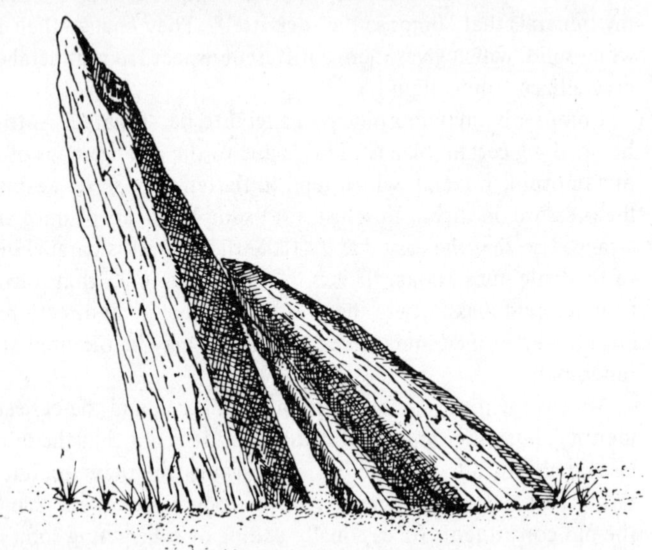

Fig. 14. "Tombstone rock," metamorphosed volcanic rock that crops out as isolated slabs in the gold country. It is the product of undersea volcanoes of 140 million years ago, now upended and changed by mountain-building forces. A field of such slabs reminded early miners of an untended cemetery – hence the name "tombstone rocks," "gravestone slate," or "gravestone schist."

The types of rocks in both groups are similar. They have been changed from layers of sand, mud, and lime to sandstone, shale, and limestone, and now, after metamorphism, they are to be seen as uneven layers of slate, schist, phyllite, hornfels, and marble.

Metamorphosed volcanic rocks are prominent among the old rocks in the gold country (see figure 14). Some of them once were "pillow basalts" (a form that lava takes when it flows into the sea or out of volcanoes on the ocean floor), in places surrounded by red chert. If you look at the chert through a microscope, you may see skeletons of the radiolaria that served as food for the larger animals in those ancient seas. The chert lenses are still recognizable, but the volcanic rocks, whether they were pillow basalts, lava flows, or volcanic ash falls, are harder to identify. Most of them have become schist.

In places in the high country, the Mesozoic and Paleozoic groups can be separated by their color when weathered. The Paleozoic ones are reddish brown on their exposed surfaces, while those of Mesozoic age have turned gray. Mixed in among or next to the old rocks are bodies of serpentine, a shiny green rock derived, perhaps, from material brought upward from the earth's mantle (see figure 15). Through the processes of mountain-building it has been altered from primordial earth material to green rock.

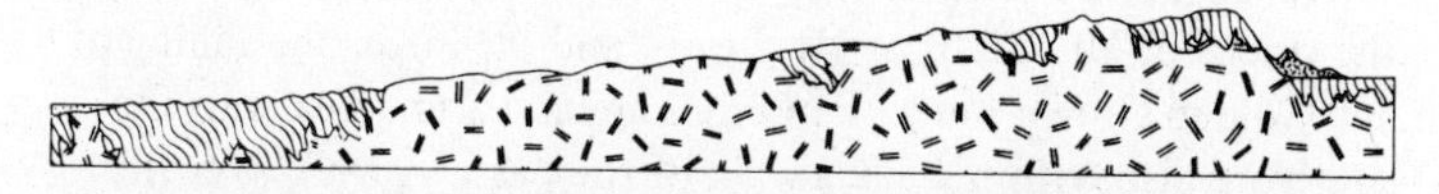

Fig. 15. Cross section showing the arrangement of igneous and metamorphic rocks in the Sierra Nevada. To the west, metamorphic rocks are to be seen along the foothills; in the range crest, you will see mostly granitic rock, but here and there a remnant of metamorphic rock remains as a reminder that the sea once covered the area where the range is now. Such isolated fragments of rock are called "roof pendants," meaning that they hang down into the rock that was under the roof of a molten magma chamber.

The best way to become familiar with any of these old rocks is to see them in the mountains – to study their outcrops. It is much more difficult to understand a rock that has been separated from its natural context. One of the most pleasant ways to see a great number of exposures of these rocks is to drive along State Highway 49 – the Mother Lode Highway – from its beginning near Mariposa to its end at State Highway 70. And the drive from Yosemite on the Priest Grade road spreads out a superb exhibit of metamorphosed rock, stacked upright for inspection. The oldest rocks in the Sierra are exhibited in the panorama at Convict Lake, near Bishop. Almost the whole Paleozoic story is told in these rocks (see color section).

The greatest thickness of older Mesozoic rocks in the high country is to be seen in the wildly beautiful reaches of the Ritter Range. From the John Muir trail northward from the Devils Postpile, climb the steep trail upward to Shadow Lake to pass through a series of volcanic and sedimentary rocks of Triassic and Jurassic age that have been metamorphosed into bands of sparkling minerals (see color section). Some minerals are fairly rare; one zone that has the manganese-rich epidote mineral, piemontite, as a prominent constituent can be followed as a reddish band for two miles through the gray schist upward and northward nearly to Thousand Island Lake at the foot of Mt. Ritter and Banner Peak.

Another hike through metamorphic rock starts at Tuolumne Meadows and goes to Mono Lake through Bloody Canyon – a principal route for Sierran traffic before the Tioga road was built. There, at Mono Pass, you may study old metamorphic rocks on the slopes of Mt. Gibbs or Mt. Lewis, and pick over the tailing of the Golden Crown mine. Parker, Koip, and Kuna Peaks to the south are also within a day's walk of the pass. A walk over the shoulder of Mt. Lewis into Parker Pass, up Koip and Kuna peaks, then back along the sheared, metamorphosed zone below Kuna Crest is a geologically revealing, if tiring, journey. Atop Parker Peak is an outcrop of dense black hornfels that breaks into smooth fragments.

Sierra Buttes, to the north, provide a less exhausting trip. They are composed of metamorphosed rock that has been called "quartz porphyry," "metarhyolite," and "keratophyre," all rocks that were exploded from undersea volcanoes 350 million years ago.

Table 2 lists other places to see examples of metamorphic rock. They are suggestions that may start you on your way to understanding the Sierra Nevada better, and will surely provide impetus for many interesting drives and hikes into the wilderness.

Table 2

Metamorphic Rocks Commonly Found in the Sierra Nevada

Kind of rock	Probably derived from	How to recognize it	Where to see a good example
Calcareous rock (limestone, marble, dolomite, calc-hornfels)	Limestone, dolomite	Generally softer than other metamorphic rocks Usually effervesces ("fizzes") in acid May contain fossils	*Caves* Clough (map 13, circle 12) Moaning (near Columbia) (map 8, circle 5) Bower (map 8, circle 10) Boyden (map 13, circle 1) Crystal (Sequoia National Park) (map 13, circle 3) Mercer's (near Murphys) (map 8, circle 1) *Quarries* Cool-Cave Valley (map 4, circle 9) San Andreas (map 7, circle 17) Diamond Springs (map 5A, circle 2) Columbia (map 8, circle 6) Volcano (map 7, circle 4)

Kind of rock	Probably derived from	How to recognize it	Where to see a good example
			Exposures from mining Columbia (map 8, circle 6) Volcano (map 7, circle 4) Volcano, buildings (map 7, circle 5) Indian Grinding Rock State Historic Park (map 7, circle 6) North San Juan, where river and State Highway 49 meet (map 2, circle 4)
Hornfels	Can be anything; calc-hornfels derived from calcareous rock	Very fine grained; may have some large crystals in fine-grained matrix Has flinty appearance	Ellery Lake (map 9, circle 8) Tioga Lake (map 9, circle 9) Mt. Lewis, Mono Pass (map 9, circle 22) Bond Pass, Emigrant Basin (map 6, circle 16) Minarets Lookout (map 9, circle 44) Mono Pass (map 9, circle 17)
Greenstone	Basalt and andesite; possibly also peridotite, gabbro, or serpentine	Very fine grained Dark Does not show individual grains or crystals Weathers to red soil	American River Canyon near bridge on road from Cool to Auburn (map 4, circle 8) Jamestown (map 8, circle 7)

Kind of rock	Probably derived from	How to recognize it	Where to see a good example
			Coulterville (map 8, circle 11) Many places along State Highway 49
Slate	Shale, tuff	Very fine grained; grains cannot be seen even if one uses a hand lens Shiny surface Breaks along flat planes ("fissile"; has "slaty cleavage")	Agua Fria quarry (map 8, circle 21) Chili Bar mine, on south side of South Fork, American River, 3½ miles north of Placerville (map 5A, circle 1) Cape Horn, State Highway 4, Dardanelles Cone 15-minute quadrangle (map 6, circle 10) Yuba River and roadcuts east of Downieville (map 2, circle 2)
Phyllite	Shale	Breaks along flat planes Individual grains visible, especially mica	Merced River Canyon, State Highway 140 at the "Geological Exhibit" (map 10, circle 8)
Schist	Shale, volcanic rock, fine-grained sandstone, shaly sandstone, chert, or any fine-grained rock	Flaky Minerals large enough to be seen; micas usually obvious In irregular, twisted layers (foliated)	"Tombstone rocks" (also called "gravestone slates") in Mother Lode area. Good field on Highway 4 west of Copperopolis (map 7, circle 23) Coulterville (map 8, circle 12)

Kind of rock	Probably derived from	How to recognize it	Where to see a good example
			Schist quarries at Mt. Ophir (map 8, circle 20) and French Mills (map 8, circle 15) Bond Pass, Emigrant Basin (map 6, circle 16) Shadow Canyon, off John Muir trail, near Devils Postpile (map 9, circle 43)
Gneiss	Sandstone, granite	Medium to coarse grained Has knotted and gnarled appearance Commonly contains quartz, feldspar, and dark minerals	Garnet Hill, Calaveras County (map 5B, circle 4) Twin Lakes, Fresno County (map 11, circle 19) Not common in Sierra Nevada
Serpentine	Peridotite, pyroxenite, other rocks composed of dark minerals	Usually fine grained, but may be granular Green to black in color Generally breaks in slivers Greasy look and feel	Bagby grade, 6 miles north of Bagby (map 8, circle 17) Rollins Lake (map 4, circle 4) Common in western foothills of Sierra Nevada
Chert (meta)	Chert; replacement in other rocks	Flinty appearance Grains not discernible Dish-shaped (conchoidal) fracture pattern	Hunter Valley, Mariposa County (map 8, circle 19) Merced River Canyon, State Highway 140 at the "Geological Exhibit" (map 10, circle 8)

Kind of rock	Probably derived from	How to recognize it	Where to see a good example
			South Fork, Wolf Creek, Nevada County (map 4, circle 6)
Quartzite	Sandstone	Fine to medium grained Generally layered Very hard May be light colored	Bond Pass, Emigrant Basin (map 6, circle 16) Convict Lake, near Mt. Morrison (map 11, circle 10) Miningtown Meadow (map 11, circle 23) and Grouse Lake (map 11, circle 24), Huntington Lake 15-minute quadrangle Not common in Sierra Nevada; most rock called "quartzite" in older reports is here called "chert" (metachert); see chert

Chapter 5

GREAT IS GRANITE

"Great is granite and Yosemite is its prophet," wrote Thomas Starr King in 1860. And surely, if one walks the high country or the deep canyons, he does get an overwhelming impression of granite – of the granite that is the very heart of the mountains.

Sierran granite – more properly, granitic rock, because you must include granite as it is technically defined, as well as all of its relatives – is part of a vast field of rock that underlies the mountains. It is exposed along the crests and extends downward an unknown distance into the earth (see figure 15). Such a field is called by geologists a "batholith," meaning "deep rock." It – or closely related batholiths – extends southward as the "Southern California Batholith" and northward into the Klamath Mountains, where the serrate peaks of little-known Castle Crags are a rock-climber's challenge.

Although the batholith is exposed in patches throughout the Sierra Nevada, the exposures do not form one huge, uniform mass of cold gray granite as one might think. Instead, the batholith is a group of individual masses or "plutons" (named for "Pluto," god of the underworld) that are distinguishable from one another. In size, the plutons range from less than a mile in diameter to more than 500 square miles. There are a few large ones that have smaller ones grouped around them. Some of the plutons are separated from one another by areas of metamorphic rock, or by thin bands of other types of igneous rock ("dikes"); others butt sharply against one another.

Most of us recognize granite as a hard, gray rock that has a salt-and-pepper appearance. A close look with a hand lens shows that the salt-and-pepper effect is created by the individual grains that make up the rock. Some grains are glassy (these are quartz); some are shiny pink, white, or gray (these are mica or feldspar); a few are black or brown (these are mica, amphibole, or pyroxene).

In older geologic reports on the Sierra Nevada, most of the

range is referred to as "granite." In modern reports, the word "granite" is scarcely to be found, although "granitic rock" is common, as are "granodiorite" and "quartz monzonite," accompanied by such strange words as "trondjhemite." The reason for this is that geologists have honed the word "granite" into a technical term, and have redefined it so that there are many more words now to cover what "granite" used to mean (and still does, to those who work in the granite industry). Figure 5 provides a simplified version of the major groups of plutonic igneous rocks.

Bear in mind that the essential feature of all plutonic rocks is that individual mineral grains can be seen by the unaided eye, and that many of the grains have sharp, straight edges.

Almost all of those who have studied the Sierra Nevada agree that the plutonic rocks of its core formed as crystals solidifying within a liquid. No one has seen granite form in the earth from fluid rock, but laboratory and field deductions have given this picture of how it may happen: The liquid itself is hot and thick – perhaps like mush – and crystals continue to form until there is no more liquid, either because it is suddenly chilled to rock or totally crystallized. There is not always room in the thick liquid for geometrically beautiful crystals to form; more often, they are pushing one another for room to grow, so that few, if any, are allowed to form into what should be their perfect shape. When crystals do have enough room and time to grow perfectly, they are greatly admired and highly prized.

By laboratory experiment and careful observation of rocks in nature, geologists have determined which minerals tend to be the first to form from the hot liquid, and which tend to be the last (see figure 16). Minerals that take longest to melt – that require the most heat – are those that form first as the hot liquid is cooled. Those that melt easily, at a low temperature, stay in solution longer while the liquid cools, precipitating only when the mass reaches the point at which these minerals would melt if it were being heated.

In general, dark-colored, heavier, calcium-, iron-, and magnesium-rich minerals form first, just as they melt last when heated. They, and some of the feldspars – those with the most

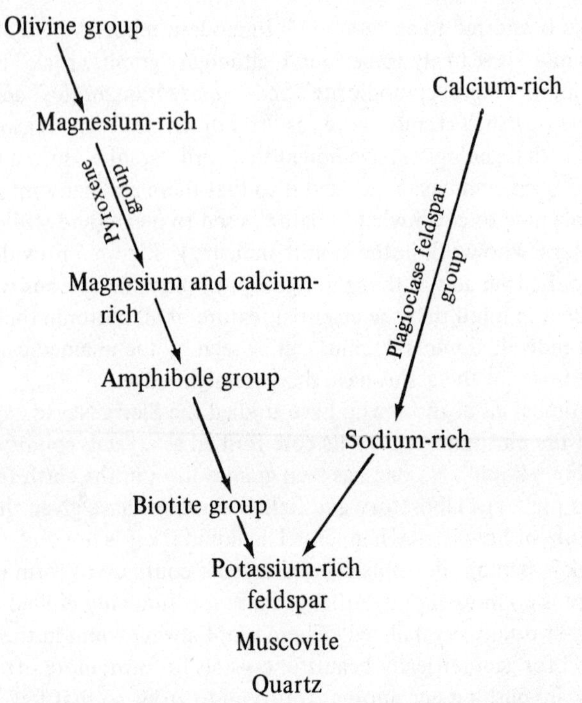

Fig. 16. The theoretical order in which groups of minerals crystallize from a hot magma. First to form are minerals rich in iron and magnesium, followed by those with calcium, then those containing appreciable sodium, and finally those having a substantial amount of potassium, but lacking magnesium. All of the minerals listed here are "silicates;" that is, they have silicon dioxide as a constituent. Quartz is pure silicon dioxide, and does not usually form until the other ingredients are used up. If the magma is disturbed, either by sloshing in the magma chamber deep within the earth, or by the addition of new magma, or by melting of older rocks, or in any other way, the course of crystallization is changed.

calcium in them – are the principal constituents of the dark side of granitic rock clan. As the melt cools, other minerals are crystallized: feldspars rich in sodium, together with the dark mica, biotite; finally, feldspar rich in potassium, followed by the clear mica, muscovite ("isinglass"), and, last of all, quartz (see figure 1).

Since most of the minerals that form before quartz contain silica (silicon dioxide – white quartz is pure silica), it follows that most other ingredients (iron, magnesium, sodium, calcium, aluminum) are used up before much quartz crystallizes, and that quartz must be the residue of the solution.

As the minerals form, they tend to settle toward the bottom of the liquid, if it is undisturbed. The reason is that the earliest minerals to form are also the heaviest – heavier than the liquid itself. The next mineral in order is lighter in weight than the first (usually in color, too), but still heavier than the liquid. It, too, sinks, making a layer on top of the heavier, darker ones already at the bottom. In theory, if the liquid is never stirred or interfered with in any way, a series of plutonic rocks should form that grades from dark, heavy rocks (such as peridotite) through intermediate gray ones (such as gabbro and diorite) to granite, the lightest in color and weight.

One of the last products of magmatic crystallization is likely to be granite in the technical sense: coarse-grained, and composed of quartz and feldspar that is rich in sodium and potassium. Most Sierran granitic rocks are not rich enough in potassium to be called granite; instead, they contain a higher percentage of sodium- and calcium-rich feldspar, and technically fall into the granodiorite or quartz monzonite group. In geologic parlance, the liquid melt from which the rocks crystallize is called a "magma;" this entire process of crystallization is called "magmatic differentiation."

One can think of many factors that might determine what kind of rocks ("differentiates") and how much of them any magma will produce. The original chemistry of the liquid mass is certainly one factor; the possibility of sloshing or stirring as the crystals cool is another; the melting of the earth's rock that encloses the liquid mass surely is a third.

The study of volcanoes has revealed that there are many dissolved gases in molten rock. As these find their way to the surface through cracks in the rock or through volcanoes, the chemistry and physics of the mass are altered, and the whole course of crystallization may change. If volcanoes on the surface of the earth throw out liquid magma from the pluton, such a process surely affects the chemical content of the remaining liquid; but how and to what extent is bound to be different for each magma and for each pluton.

Although the principles that govern the history of magmas are the same for all plutons, each pluton is an individual, and has an individual history.

In many of the granitic areas of the Sierra, the rock is spotted. On close inspection, the spots prove to be patches of either metamorphic rock or of darker igneous rock. Certainly there must have been many pieces of the older metamorphic rock torn off to be melted in the hot magma. The fragments left for us to see are those that never were fully digested – either because they were the core of a larger mass that did not get completely melted, or because they were pulled into or surrounded by the magma when it was too cold to dissolve them.

In some places of the world, there is evidence that has led some geologists to the conclusion that granitic rocks are not necessarily formed from a hot melt. They reason that granite may, instead, be formed cold, where it is, by a reorganization of its chemistry, aided by chemical ions migrating through the atoms of the rock. However, no one working in the Sierra Nevada has yet suggested a cold origin for its granitic rock.

On the other hand, evidence supporting the idea that Sierran rocks cooled from a hot body is fairly convincing. One such piece of evidence is this: a close look along the edges of many of the plutons, will show that the rock is finer grained there than in the center. This is what would be expected of crystals forming in a cooling liquid; the slower the liquid cools, the larger the crystals grow. If the hot liquid meets the cold rocks of the rest of the earth, as it does along its outside edges, it is cooled there, and whatever size the crystals have reached when the edges solidify is the size they will be. In the center, where the liquid did not meet the chilling edge, the crystals can go on growing until the whole pluton becomes solid rock – unless something disturbs the magma.

The Sierran batholith is far from being perfectly understood.

Now that we can tell time backward by radioactive clocks, it is possible to begin to unravel the batholithic history. Researchers who have analyzed potassium:argon ratios of the many plutons that comprise the batholith have suggested that magma invaded the Sierra in distinct pulses. Each pulse lasted from 10 to 15 million years, and was separated from the following one by about 30 million years. Altogether, it took 130 million years to complete the creation of the granitic rock.

Figure 17 shows rocks that cooled in these pulses as they are now understood. All California plutons are not shown on this tiny map – not even crudely. The Southern California Batholith is missing, although it is surely related. Only a few of the plutons in Nevada and eastern California are shown, although they, too, are a part of the story. Nor are the Klamath Mountains shown, though they probably should be considered as an extension of the Sierra Nevada, separated from the main mass by a volcanic cover.

Whether or not the pulses of igneous activity indicated on this map are pulses in a rhythmic sense is not known. Perhaps they are periodic, as this crude arrangement makes them appear. If so, one cycle of granitic emplacement should have taken place in the Tertiary Period, and another could be underway now. Since such an event would be taking place far beneath our feet, we do not now have a way of knowing if the periodicity is real or not.

It is granitic rock as a whole that determines the shape of the mountains. It is the response of granite – the whole family of granite – to the ravages of weather, to glaciers, to streams that come down the mountains that has given us the grandeur of today's Sierra Nevada.

Two characteristics of granitic rock that have been responsible for much of the spectacular scenery of the Sierra are jointing and sheeting, both expressions of the way in which rock breaks. "Joints" are more or less even planes along which the rock cracks – generally up and down as well as in two horizontal directions. "Sheeting" describes the cracking of a rock along curved surfaces parallel to the surface of the rock. Jointing makes sharp, steep faces like the east side of Mt. Whitney; sheeting provides the magnificent domes of Yosemite (see figures 18 and 19).

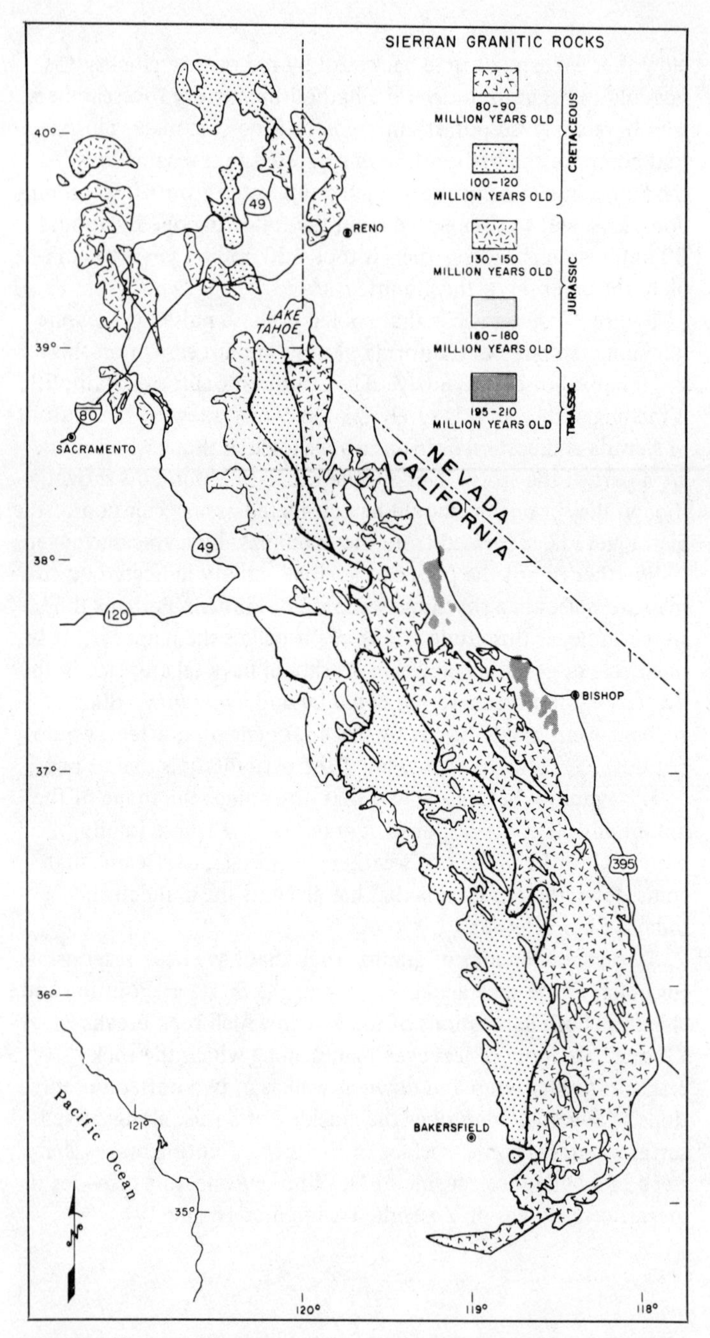

Fig. 17. Groups of Sierran granitic rock that formed during several stage of cooling as interpreted from the actual ages of granitic rocks.

Fig. 18. Splintery Sierran peaks that owe their shape to prominent vertical joints.

Fig. 19. Rounded dome on the Tuolumne River. Domes like this and many others in the Yosemite region remain as monoliths because they are relatively unjointed and resist weathering.

Joints are regional features, cutting across many miles and through the entire batholith. They may be identified as a faint criss-cross pattern on aerial photographs, or they may be obvious in individual outcrops, where they divide granitic rock into blocks like loaves. Such a heap of loaves is seen along State Highway

108, the Sonora Pass road, at Eagle Creek near Dardanelle. A very prominent joint has cut a conspicuous slot in the mountains on the High Sierra trail near Hamilton Lakes.

On State Highways 88 and 89, in Alpine County, and on U.S. Highway 395 north of Markleville, the mountains have ribs of granitic rock, left after erosion along adjacent joints wore down the intervening rock. In wet weather, water courses down between the ribs, etching them more deeply. Throughout the Sierra, joints influence where streams run, noticeable especially in the Alabama Hills on the east side of the Sierra.

In the Sierra there are two major sets of joints, one trending northwest with the grain of the range, the other northeast at right angles to it. Here and there they change direction slightly, but they change together so as to maintain their nearly perpendicular relationship. Where the rocks are fine grained, joints are close together; where they are coarse grained, farther apart. They pass from one pluton to another almost without deflection. Although the overall pattern is a general criss-cross, the individual joints are short, several miles being the greatest length of any single one. That this is the regional pattern indicates that it developed after the batholith crystallized, or was impressed upon the batholith as it cooled. It does not seem to be a direct result of cooling (there are cooling fractures in small bodies of igneous rock), because the masses cooled at different times. On the other hand, the pattern developed before the extensive weathering period that followed (see Chapter 7), because the joints themselves are deeply weathered.

How and why did it develop? No one as yet knows.

Sheeting, in contrast, is a local phenomenon. It can be observed on an individual outcrop, on a single mountain, or on a particular dome, but there is no regional sheeting pattern. The sheets form parallel to the topography, and, where they are nearly straight, may resemble joints. It is to this tendency of granitic rock to form sheets that Half Dome, Sugarloaf, Liberty Cap, Pywiack, Lembert, Fairview, Polly, and the many other Yosemite domes owe their remarkable shape.

Unlike the sharp, ragged peaks of jointed rock, such as those in the High Sierra near Mt. Whitney, the curious domes of Yosemite are usually formed in less jointed rock. They are the products of exfoliation – "leafing away" of layers of granite much as layers of onion peel away from the center. The original form probably was rectangular, or at least much more angular, but through gradual exfoliation the sharper edges have dropped away (figure 20).

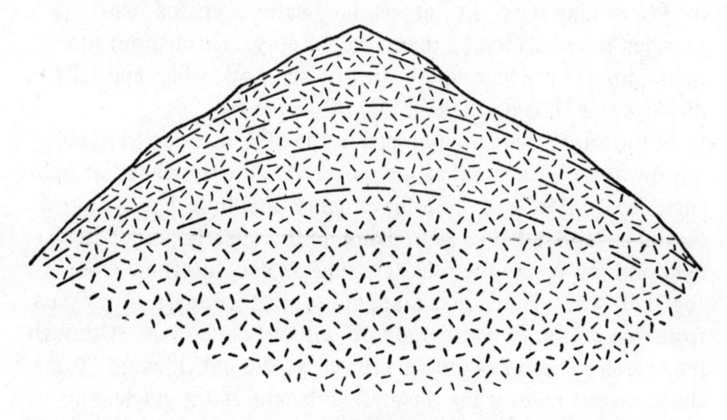

Fig. 20. How weathering wears off angularities, leaving a rounded boulder.

Just why the rock should leaf away is also somewhat of a mystery. Heating by the sun has been suggested as the principal cause. The sun heats and expands the rock, then, when it cools and contracts (in winter, freezing), a split begins. But this cannot be the whole story; if it were, the process should operate very rapidly in the desert, where there is more sun and more heat. Yet it does not seem to.

Laboratory experiments that simulated nearly a thousand years of alternate heating and cooling had very little effect on granite. On the other hand, heating and cooling combined with simulated rain began to produce shells of exfoliation within two and a half simulated years. The reason for this seems clear enough: feldspar minerals, when water is added, form clay, which not only crumbles away when dry but swells when wet, shouldering other minerals aside. It is certainly true that the rounded tops of exposed granite domes do not easily break into sheets, whereas the sides do. In contrast, domes or "corestones" buried in the ground weather all around into shells.

It is also true that feldspar crystals are longer in one direction than in the other two; this will help to orient the direction of peeling. But true circular weathering is hard to account for completely. Perhaps, as one geologist has suggested, the original orientation of minerals in the cooling magma has partly determined that rounded forms will develop.

Sierran granitic rock breaks into sheets in many places where it does not produce rounded domes. Sheeting of granite in Big Arroyo, High Sierra, gives the mountainside a peeling aspect; along Tioga road near Yosemite Creek campground in Yosemite, and at Tragedy Springs, on State Highway 88, sheets of granitic rock are stacked like bricks, as if a building had been intended and forgotten.

It is not hard to see granitic rocks in the Sierra Nevada. Every major pass through the mountains exposes acres of gray granite as it cuts through the mountain core. The map on p. 70 shows the extent of these bodies in the Sierra, divided by age rather than rock type. Each major intrusive epoch produced granitic rocks of all sorts, so that one can see granite and its relatives that are as old as 210 million years, or as young as 80 million years – all part of the great Sierra batholith.

The best known granitic rocks in the Sierra are the spectacular ones in Yosemite National Park. Frank C. Calkins, who studied these rocks more than forty years ago, has made a superb map of much of the park, which can be used as a guide to the rocks. On his map, he has shown where to find the striking Cathedral Peak granite – that whitish gray rock with the huge feldspar crystals; where to see gabbro and diorite; what Half Dome, Sentinel Rock and Mt. Clark are made of, as well as a great deal more.

A hike in the Huntington Lake area, east of Fresno, will reveal several different varieties and colors of granitic rock. There is the lighter-colored quartz monzonite of Bald Mountain; the grayer granodiorite of Dinkey Creek, and the light gray quartz monzonite of Dinkey Dome, which here and there contains smoky quartz crystals.

Granite is not as popular for building stone as it once was, since newer architectural materials have been developed that are cheaper and safer. But the beauty of polished granite is enough to encourage some architects to use it, despite the cost. It is still shaped into monuments, although the use of monuments, too, is declining.

At one time, several quarries in the Sierra Nevada supplied stone for cities throughout California. In San Francisco, the dark granitic rock with purplish quartz and orange feldspar in the Hibernia Bank building came from Rocklin; the Bank of California, on Sansome and California Streets, the St. Francis Hotel, the Dewey Monument in Union Square, the Old Custom House on Battery Street, and the new Post Office on 7th and Mission Streets are of stone from the Raymond quarries. In Sacramento, the Capitol building has two types of commercial granite: the base is of Penryn granite; the upper part is 130-million-year-old Rocklin granite. The Sacramento City Hall has steps of Rocklin rock, as does the Cathedral of the Blessed Sacrament. The U.S. Post Office has Rocklin granite walls.

In Los Angeles, the Fountain Mall in Civic Center features one large piece of Raymond granite. Although much granite in Los Angeles came from quarries in the Peninsular and Transverse Ranges, the Federal Reserve Bank contains Sierran granite.

Most granite quarries are on the western side of the Sierra, partly because it is closer to large city markets, and partly because masses of granitic rock on the western side are less jointed than those on the east. The domes of Yosemite, for example, hold together through time far better than the more jointed granite in the vertical gothic rock spires on the eastern side.

Partly because of the way the east side granite breaks, and partly because the east side is not so well clad with trees as the west, it is an excellent place to see many different types of granitic rocks. Geologist Paul Bateman, who has worked many years in the high country west of Bishop, has suggested that from the outskirts of that town under the morning sun one can see several different bodies of granite in one view. Afternoon shadows obscure the distinction between the light gray granite of Mt. Tom, the still lighter, younger granite in Mt. Emerson, and the older, darker granite to the south.

Or, walking up the trail from Glacier Lodge west of Big Pine to Palisade glacier, you can see along the trail granitic rock of all sorts and colors. At the end of the trail, there is a view, not only of granite, but also of the largest of California's remaining glaciers.

Quite a different aspect is presented by the granite of the Alabama Hills, at the foot of the Sierra near Lone Pine. Reputed to be the "oldest" rocks in California (quite untrue), they have a tangled, gnarled look, as if they were weary of what they have endured. They are rounded and deeply weathered; it is easy to see how they gained their undeserved reputation for great age.

A careful look at the Alabama Hills reveals that these rocks, too, are jointed and faulted. Yet their aspect is quite different from the faulted, jointed granite of the Whitney area above them. In both regions, the granite is about the same age: 80 million years. What could account for the difference in appearance? Climate? Surely the desert climate will shape a far different character than the high, cold mountains. What difference will the elevation make? And what else will change them? What has been their story through the ages? Have they been glaciated? These are questions we are all free to speculate upon, and to test our speculations as best we can.

Perched against the backdrop of the highest part of alpine Sierra, the Alabama Hills are a lure to photographers. In fact, so many "Westerns" were made on location in the Alabama Hills that there is now a scenic drive marked through the hills. Moviegoers may recognize Movie Flat, a ranch that has seen many a shootout, the rocks, the road, and the mountains. There is also a large native Indian population in the eastern Sierra, who were often actors in these dramas.

Table 3

Plutonic Igneous Rocks Commonly Found in the Sierra Nevada

Kind of Rock	How to recognize it	Where to see a good example
Granite (includes granodiorite and quartz monzonite)	Grains large enough to distinguish from one another Contains quartz, orthoclase feldspar, and potassium feldspar; may contain some dark minerals Usually has salt-and-pepper appearance; may be pinkish	Yosemite National Park (map 10) Alabama Hills (map 14, circle 2) Mt. Whitney (map 14, circle 4) Desolation Valley (map 3, circle 8) Indian grinding rocks, at Yosemite Valley (map 10, circle 16) and Grover Hot Springs State Park (map 6, circle 4)
Diorite	Grains large enough to distinguish from one another No quartz; plagioclase feldspar and dark minerals comprise most of the rock (dark minerals are less than 50%) Gray	El Portal (map 10, circle 9) The Rockslides, Yosemite National Park (map 10, circle 11) Interspersed with metamorphic rock near Copperopolis (map 7, circle 22) In smaller plutons in the western Sierra
Gabbro	Grains large enough to distinguish from one another No quartz; plagioclase and dark minerals comprise most of the rock (dark minerals are more than 50%) Dark gray to black	1½ miles north of Camptonville on State Highway 49 (map 2, circle 3) In roadcuts 6-8 miles west of Hornitos (Guadelupe intrusive complex)(map 8, circle 16) Near Rough and Ready, Nevada County (map 4, circle 2) Twin Lakes, Fresno County (map 11, circle 19)

Kind of Rock	How to recognize it	Where to see a good example
Peridotite	Grains large enough to distinguish from one another Dark green or black No quartz, no feldspar Dark minerals only	Red Hill, northwest of Meadow Valley, Feather River, Plumas County (map 1, circle 1) Pulga, Butte County (map 1, circle 3) Most peridotite in the Sierra Nevada has been partly or wholly altered to serpentine
Porphyry	Lumpy; some crystals much larger than others	Cathedral Peak, Yosemite National Park (map 10, circle 4) Lembert Dome, Tioga Pass road, Yosemite National Park (map 9, circle 13) Both of these are porphyritic granite exposures; better example of porphyry is the rhyolite dike west of Convict Creek, Mt. Morrison quadrangle (map 11, circle 10)
Pegmatite	Very large crystals, usually quartz, feldspar, and mica	Kingsbury grade, at sharp curve where road becomes less steep (map 3, circle 4)

Chapter 6

TREASURES FROM THE EARTH

Compared with the rocks of the earth, deposits of metals are very scarce. They are oddities of nature, and only man's fascination with them has transformed them into treasure.

Gold, for example, is found in the sea as well as in most rocks. In igneous rocks, the amount of gold averages about five ten-millionths of one percent (0.0000005%). Most gold mines of economic value – that is, those that pay the cost of mining – are in areas where nature has concentrated the precious metal twenty thousand times that much, yet the ore contains only one-third of an ounce of gold per ton of rock! Even today, when the cost of mining is very high, huge floating dredges, similar to those so common on California rivers until a few years ago, can make a profit mining ore that contains only five cents' worth of gold per cubic yard of gravel. This is a concentration of about 1/800 of an ounce per yard, or about one part in 32 million.

It is this rarity, coupled with the fact that gold is found almost everywhere in some amount, that has led to the idea that "gold is where you find it." Indeed it is; but gold in commercial quantities, like most other metals, is by no means evenly scattered throughout the earth. Gold concentrated by nature in quantities large enough to mine is found in favored regions – generally in mountains.

The Sierra Nevada is one of those favored regions, where an amazing amount of gold has been won from a small area. Some of it was in enormously rich zones called "bonanzas." For example, one single lump of gold taken from the Carson Hill mine near Melones in 1854 weighed 2340 troy ounces (160 pounds avoirdupois), and was then worth nearly $44,000!

The forces that controlled the building of the Sierra twisted and bent the layers of ancient Paleozoic and Mesozoic rocks several times, indicating that there were several episodes of mountain creation before the present Sierra took form. During this mountain making, particularly during the later episodes in the Jurassic Period (about 150 million years ago), the way was prepared for gold ore to form, for then the Mother Lode fault zone and many of the numerous related faults were born. These great tears in the earth must have been the source of many earthquakes a hundred million years ago. Cardinal to the making of mountains, the faults opened avenues along which mineralizing solutions could rise from the depths. Today, the old faults are healed by quartz and other minerals – including gold and silver; it is very doubtful, but not impossible, that they will again be "earthquake faults" in our lifetime.

The gold and silver, as well as many of the other ores, were formed by the grace of granite. For it was during the cooling of the granite magma that hot waters and gases steamed upward, penetrating the rock through joints and fractures, leaving behind telltale evidence of their passage. Nowhere was the evidence more apparent than near the fault breaks, where the mineralizing solutions left veins (thin, wandering sheets, not cylindrical tubes like blood vessels) and "vugs" (holes) filled with ore-bearing minerals, as well as metal shot through the rock on either side. The ore-bearing solutions were not particular as to what sort of rock they left gold deposits within; nearly all of the old metamorphic rock types contain gold deposits somewhere in Sierran gold country.

It is not true that all gold is bound in quartz, or that all quartz veins contain gold. Several generations of Sierran quartz mark different episodes of solution escape, but the gold-bearing quartz was usually the last to form. Where the veins turned, or where they split or swelled, proved to be places where gold accumulated. Experienced miners in the Alleghany district of Sierra County claimed that they could tell "live" quartz that might contain gold from "dead" quartz that never did, by its color: "live" quartz was milkier, less lustrous.

Perhaps, indeed, they could; Henry Ferguson, a geologist who studied the veins in the 1930s, thought so. Although the microscope showed him that the "live" quartz was microscopically broken, he himself never gained enough experience to separate "live" from "dead" in the mine. What's more, the miners, even though they could recognize "live" quartz that might contain gold, could not tell if it actually did or not. They could only separate out the kind that definitely had no gold – the "dead" quartz.

Many other puzzles about the gold and the quartz veins are still unsolved. How far down do the veins go? Judging by chemical and physical evidence, they were formed deep within the earth, probably more than 10,000 feet below the surface. Mines of the Mother Lode follow them for more than a vertical mile, yet there is no evidence of any change in them, or of their beginnings.

In some places in the gold country the quartz veins have openings in which quartz crystals have grown large enough to warrant mining for use in radio equipment. One can understand quartz crystals growing large and splendid where there is room for them. But how were the strong rocks pushed aside enough for quartz crystals to grow into veins? Surely deep within the earth there are no real openings – merely planes of weakness. It does not seem likely that quartz, solidifying from a hot solution, could exert enough strength to force the host rocks apart as far as the width of the vein. Neither does it seem likely that the solutions had wide avenues through which to rush unimpeded. Rather, they must have moved in molecule by molecule, in places shoving aside an earlier molecule, in other places filling a minute void.

Although it is true that faults served to determine where ore deposits would be, there are areas in the gold country, particularly in the region of the northern mines, where the main fault

Fig. 21. Forty-niner.

zone is masked by a belt of serpentine and related rocks. Such rocks – the peridotite group as shown in the chart on p. 14 – may have come from deeper within the earth, instead of separating out from magma that eventually formed granite.

The peridotite group is host for many mineral deposits. Asbestos fibers that have grown within the serpentine were mined near Copperopolis; chromite deposits (from which chrome is derived) are scattered throughout the heavy, dark rocks, and are now and then mined, depending upon our national need; magnesite, the gem stones chrysoprase and idocrase, and the striking green mica, mariposite, are in rocks related to the serpentine belt.

In the 1960s, deposits of nephrite jade were recognized in serpentine in Mariposa County, not far from Yosemite National Park. Since these deposits are in the heart of the gold country, one may wonder if the Chinese miners of yesteryear overlooked them, did not recognize them, or, since they valued jade more highly than gold, simply mined and did not tell.

There have never been extensive iron mines in the Sierra Nevada, but there are reserves of iron that have not been exploited. Most of them are too remote and in bodies too small to be economically valuable now. So long as iron can be mined more cheaply elsewhere, or can be recovered from the dumps that threaten to overwhelm us, Sierran iron should not be called upon.

One small deposit, high on the slopes of the beautiful Minarets, is contained in flat-lying lenses in metamorphosed volcanic rock. The ore is magnetite – one of the few minerals that can be identified by noting its pull on the needle of a compass. Hikers exploring the sharp Minarets (although the ore bodies are flat-lying, the mountains are exceedingly steep) should remember that their compasses may be deflected by the magnetite to give them false directions.

Evidence that deposits of the ores of many metals are derived from hot mineral waters comes directly from hot springs. Modern hot springs, and, by inference, ancient ones as well, deposit mercury, sulfur, and even gold as they bubble out on the surface of the earth. Steamboat Springs, a spa at the foot of the Sierra in Nevada, is one of those known to be depositing silver and gold today, in very small quantities. It probably takes its origin not from the heat of the now-cooled Sierran granitic core, but rather from the more recent volcanic source along the eastern Sierra.

In some places where molten granite has come in contact with the old metamorphic rocks, a host of "contact metamorphic" minerals have resulted. Many of them are interesting to collectors; some are rare and beautiful; a few are in sufficient quantity and of sufficient value to constitute ore. Among these contact deposits are ores of the metals tungsten and molybdenum, formed along the junction of granitic and calcareous rock – granite and limestone. The largest tungsten mine in the world is in the Sierra, 11,000 feet high in the mountains up Pine Creek out of Bishop. Both molybdenum and tungsten, used especially to harden steel, are derived from this mine, set in some of the most spectacular of Sierran scenery. Ore from the Pine Creek and adjacent mines is found mixed with garnet. Most of the garnet is pale to reddish brown; it has been sold as an abrasive, but is not gem quality. Other tungsten mines and prospects contain garnet also, and there are such superior collecting localities in the Sierra as Garnet Hill in Calaveras County.

The number of deserted mines, shafts, and prospects pits in the Sierra is uncounted. Throughout the mountains one may find open holes, caving tunnels, and ruined headframes. Although by law these should be fenced or boarded, very few are. Exploring old mines is extremely dangerous. You cannot be sure that the interior is not caved, that the air is not poisonous, that there are no internal shafts – winzes – to fall into, or, for that matter, how much the workings may wander and branch. Even if you are reasonably certain that the workings are safe, you should never ven-

Fig. 22. Adit and ore car.

ture in without wearing a hard safety hat and carrying a light, and without leaving word as to where you have gone.

It is sometimes possible to obtain permission to visit an operating mine, but so few underground mines are operating in the Sierra that the opportunity is not great.

In the gold country, although the thousands of miles of underground burrows are no longer available to us, the surface workings, ruined though they may be, are interesting in themselves. One of the most unusual is the Kennedy mine, near Jackson, which has been preserved as a park (see figure 23). For many years, the Kennedy held the title of the deepest mine in North America (Homestake, in South Dakota, is now deeper). Its workings explored the earth more than a mile vertically below the surface, along a 150-mile network.

Gold in the Kennedy was contained in a quartz vein emplaced along a fault. The vein is visible in places on the surface along Highway 49, but is unobtrusive and hard to find. As it dips eastward into the ground, it becomes more interesting, for from this branching, lensing quartz vein and its extensions, flanked by slate and greenstone, Kennedy miners have in the past taken nearly $60 million dollars in gold, worth three or four times that much at today's fluctuating prices.

Underground, the workings connect with those of the Argonaut, whose headframe may be seen on the hill to the west. It, too, is a mile-deep mine, with eight miles of drifts and tunnels, four miles of raises, and fifty miles of stope floors. In 1922, the Argonaut caught fire, killing forty-seven miners – a whole shift – while the nation waited and worried. The mine produced about $25 million worth of gold (when gold was $20.67 an ounce) before being shut down, like the Kennedy, in 1942. [1]

[1] A special presidential order closed all American gold mines during World War II. Since then, the mines have flooded with water and their workings have deteriorated. The high cost of labor and material today has made it too expensive to rehabilitate them. Even if it were economically feasible to repair them to mine gold again – there is certainly gold left – it would be difficult to find experienced miners. It is essentially a lost art in this country; most engineers and miners with knowledge and experience in Sierran gold fields are octogenarians, or close to it.

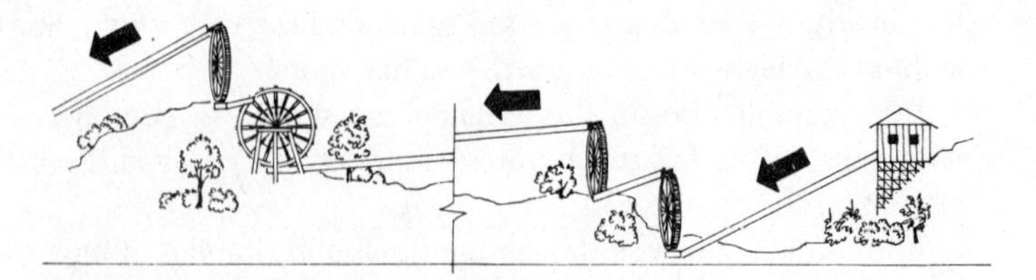

Fig. 23. How waste from the Kennedy mine and mill was lifted uphill by means of the "Kennedy Wheels" to the tailing dump. Wet waste pushed into the chute from the mill at right slid downhill to the bottom of the first wheel, where buckets built into the perimeter of the wheel lifted it to the top. From there, it slid downhill to the second wheel, where it was again lifted. Each time, one of the four wheels lifted it in elevation, until it was finally high enough to pour over a hill into the tailing dump (left). Not to scale.

Part of the surface equipment of the Kennedy mine consisted of an unusual system of wooden wheels used to carry the waste rock – called tailing – from the mill (where ore was crushed and gold separated) near the mine entrance to the dump. The wheels are now picturesque ruins to be seen south of the mine along Jackson Gate road. From a hill to the north of the road, one can see two of the pine wheels, and beyond, the headframe and remaining buildings of the Kennedy. On the skyline, the Argonaut headframe is visible.

On the hill to the south are the two remaining wheels – one of them reconstructed – and over the hill are the old tailing dumps. The wheels, when working, carried the tailing in the manner shown in the accompanying sketch. Wet tailing from the mill, lifted to the top of the mill building, was pushed into a trough or "flume" down which it flowed toward the first wheel, where it was dropped into a well at the bottom of the wheel. The first wheel, like the other three, was equipped with 176 little redwood wells, or buckets, along its perimeter. Each bucket picked up its share of wet tailing, and as the wheel was rotated by the electric motor and belt drive that powered it, the tailing was lifted to the top of the circle. There it was dumped into another flume, angled slightly downhill toward the bottom of the second, higher wheel. From there, it was again lifted to the top, where it was dumped to slide down toward the third, and similarly, to the fourth, highest wheel. Each wheel was 68 feet in diameter, and lifted the waste 48 feet vertically. Of course, each wheeel actually lifted it 68 feet, but since part of the elevation gained by the preceding wheel was lost in dumping the waste in the well, and part was lost in scooting it down the flume to the next wheel, the total amount gained was 48 feet.

From the top of the fourth wheel, the ground-up rock slid down a series of long flumes to a dump, where it grew into a flat-topped, artificial hill.

When the wheels were in operation, they were housed in sheet-metal buildings. The buildings are no longer there, and only the crumbling relics remain.

The purpose of this complicated enterprise was to protect the environment. In order to keep the tailing from entering the local water supply, this unusual artifice was developed. It is ironic that the tailing heaps have been turned today into dams for water supply!

The entire vista is far more pastoral than it was a half-century ago. Then, the one-hundred heavy stamps of the Kennedy mill, used for crushing the ore small enough to recover the gold, added to the sixty at the Argonaut, must have roared in a way no freeway yet in the Sierra can equal. Noise pollution has been abated, but so has gold mining. There are no stamp mills operating commercially in the Sierra now, but there are remnants; one small, now-silent mill, rescued from the Golden Center mine, stands in the historical exhibit on Wolf Creek in Grass Valley. It and a renovated Pelton water wheel, used to generate power for the North Star mill, are exhibited outside; inside the old power house are other mementos of mining days gone by.

Table 4

Exhibits of Mines and Mining Equipment

Where	What
Columbia State Historic Park Columbia, Tuolumne County	Underground mine open to visitors
Placerville, El Dorado County	Underground mine open to visitors
Gold Run Roadside Rest Gold Run, U.S. Interstate 80, Placer County	Hydraulic mining pit with gravel bed exposed (Tertiary river)
Columbia State Historic Park Columbia, Tuolumne County	Rocky bed of ancient river exposed by hydraulic mining
Malakoff Diggins State Historic Park North Bloomfield, Nevada County	Hydraulic mine, townsite, mining equipment
North Star Powerhouse State Historic Park Grass Valley, Nevada County	Stamp mill Pelton water wheel, mining equipment
Columbia State Historic Park Columbia, Tuolumne County	Townsite, historic equipment, scale-model hydraulic monitor
Marshall Gold Discovery State Historic Park Coloma, El Dorado County	Townsite, gold discovery site, mining equipment (especially placer)
Plumas-Eureka State Park Johnsville Star Route Blairsden, Plumas County	Stamp mill, tramway, mining equipment
Kennedy Wheels Jackson Gate Jackson, Amador County	Tailing wheels, headframe

Chapter 7

RIVERS OF YESTERDAY

About 130 million years ago, in the early part of the Cretaceous Period after the gold veins had been formed, the Sierra Nevada entered a phase of deep erosion. At that time, the western sea had its shores within the Great Valley, as attested by remnants of beach and offshore mud now turned to stone.

Until recently, geologists have considered that there was a break in our record of Sierran history at this point (an "unconformity"), marking a period in which the Sierra and its cooled granitic core were steadily eroded. New evidence, including the age of granitic rocks in actual years, points to a more complex story: even as erosion was stripping the mountains to carry their fragments to the western sea, parts of the granitic core were still hot, perhaps still moving upward. Some parts may still be rising.

At any rate, erosion during the Cretaceous and part of the early Tertiary Periods wore the mountains down – down far enough to lay bare the tops of the gold-bearing quartz veins formed deep within the range. Several geologists have tried to estimate the amount of rock torn from the mountains, using various types of calculations. Although they do not agree exactly, all suggest that nine vertical miles or more of rock was removed during this 25-million-year erosive interval, or half a foot to a foot-and-a-half per thousand years. (Compare this with the present rates given on p. 48.)

By Cenozoic time, commencing about 65 million years ago, the Sierran landscape was quite different from the deep oceans of the Paleozoic or the mountain fastness of today. A shallow, lagoon-margined sea lapped quietly against the foothills of a much lower range. The lagoons have left their mark in the remnant clay, white quartz sand, and coal beds to be seen in the foothills along the edges of the Great Valley, especially near the towns of Ione and Buena Vista. A substantial clay industry still uses clay laid down in those still lagoons.

Lignite coal found near Buena Vista, Ione, and Carbondale reminds us that the lagoons were probably lined with trees and plants – perhaps even choked with them – enough to form thick mats of dead and dying vegetation that turned from peat to soft coal, rather than mixing with mineral fragments to become soil. There is no longer a California market for soft coal as a fuel, but the mines are sporadically worked to recover a very hard commercial wax.

Judging by the fossil remains of plants found in the old gravel beds of the rivers that fed the sea, the climate was subtropical, similar to that of Orizaba, Mexico, today. Temperature on land probably averaged 65°F., somewhat cooler than the adjacent 70°F. sea. Frost was rare in lower elevations; 60 inches of rain fell in the warm season on the dense vegetation near the coast, perhaps as much as 80 inches in the uplands.

This estimate of the climate was made by paleobotanists who studied the remains of fossil trees, shrubs, and vines found in mining pits. For more than a century, at the Chalk Bluffs hydraulic gold mine, near You Bet, Nevada County, and at other places along the course of the Yuba River as it was in Tertiary time, it has been possible to find fossil plant remains. That is, it was possible. One can no longer find fossils in some of the better known localities, such as Remington Hill, for thoughtless collectors, many of whom have neither use for nor more than passing interest in their acquisitions, have removed them all.

Those fossils that were collected for their scientific use have given us a picture of the woods of those days. Although most plants that became fossils were those that lived in lowlands, where they were quickly buried and therefore have been preserved, some idea of the forests in the hills can be derived from the winged seeds and more durable leaves that floated into the river beds.

Along the high river banks, red gums (liquidambar), laurels, figs, and woody climbers grew luxuriantly among the oaks and magnolias. Higher in the hills, oaks, laurels, witch hazel, small palms, cycads, hickorys, and persimmons grew, but no conifers, indicating that this was a humid upland forest of broad-leafed hardwoods.

If the underlying rock in such a tropical forest is greenstone, serpentine, peridotite, or one of their relatives, a peculiar bright red clay is likely to form as soil. This clay, called "laterite" (from the Latin word for "brick"), is rich in iron and sometimes nickel – rich enough in nickel in some places that it is an ore. Up and down the Sierran foothills, one can see this startling red soil (colored red by the iron) – a reminder of the warmer, moister tropical climates of yesterday. At Camptonville and at Ione, the color seems unusually bright; those who walk in it, or have small children who do, remember it long and sadly at washday.

The Sierra was much lower in Tertiary times than it is today. Close to the shores of the subtropical sea, a belt of resistant greenstone ridges rose to elevations of 1000 to 1800 feet. Rivers – not where they are today – cut through the ridges in several places, flowing through narrow canyons. East of the greenstone belt, the area that is now the gold country was a flatland underlain by upended slate beds; into it the valleys of the Tertiary rivers were cut. These river valleys were broad, with the main course of the river in the center of the valley, its channel incised into fresh bedrock as much as 40 feet below the general valley level.

From the plain, the hills rose gently eastward, reaching an altitude of as much as 3000 feet along the crest. A few isolated peaks stood out from the rest: Mt. Dana, in Yosemite, for one, and Pyramid Peak for another.

Perhaps there were land birds living in those old forests, and water birds wading in the mud flats along the streams and in the ponds and oxbows. There must have been insects humming in trees, and earthworms and snails burrowing in the soil. But we have little record of them here. The Sierra Nevada is not a fossiliferous mountain range.

What Tertiary fossils we do have, aside from the plant fossils, are chiefly remnants of teeth and bones of larger animals, browsers and grazers in the savannah. Near Knights Ferry, close to Tuolumne Table Mountain, paleontologists have unearthed the remains of two little horses (*Nannipus* and *Hipparion*), two camels, a pronghorn antelope, and one tooth of a mastodon.

Nearby, at Oakdale, fragments of two ground squirrels and a rabbit, two other small horses (*Neohipparion* and *Pliohippus*), a small camel, and another pronghorn were buried. Since there

mud.

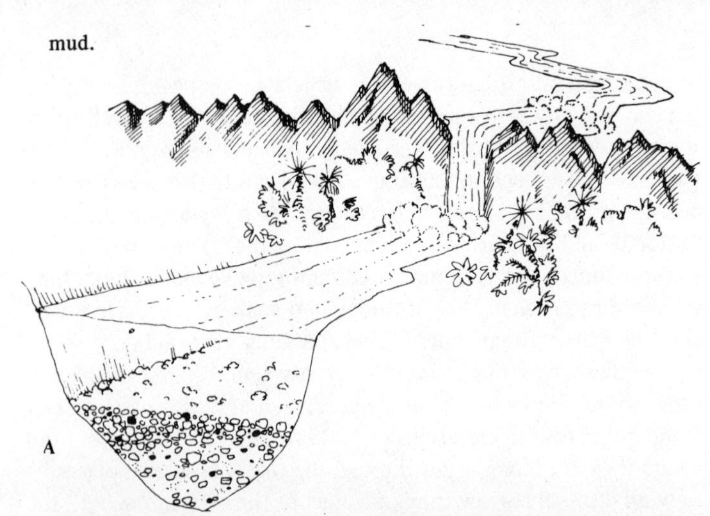

A, a stream of Early Tertiary time (about 50 million years ago) as it meanders through the gentle western slope of the Sierra Nevada. In the lower foothills, it encounters a ridge of resistant greenstone; it has cut through the ridge to plunge over a waterfall to the region underlain by soft rock below and to the west. A cut out of the stream channel shows gravel in the bottom of the bed. Mixed with the gravel are nuggets and tiny fragments of gold, worn from the higher mountains and being carried by the stream toward the sea.

B, about 30 million years ago, shows how ash, falling from volcanoes erupting higher in the mountains, has clogged the stream. Where once the stream poured over the greenstone ridge in a rushing waterfall, a dam of ash has ponded the water behind the ridge. In the cut out, ash may be seen covering the gravel of the river bottom.

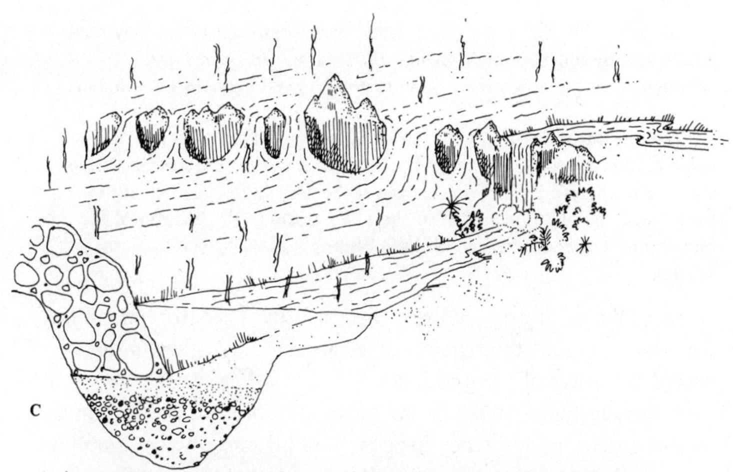

C, steaming volcanic mud flows of 20 to 10 million years ago rolling downhill from near the Sierran crest, covering much of the landscape. Here, the stream has been forced to seek a new route through the greenstone ridge; parts of the ridge and about half of the stream bed in the foreground have been covered by the mud.

The cut-out shows a succession of deposits, including the original gravel in the bottom of the channel, followed in turn by ash, then partly covered by a mixture of volcanic mud and rock.

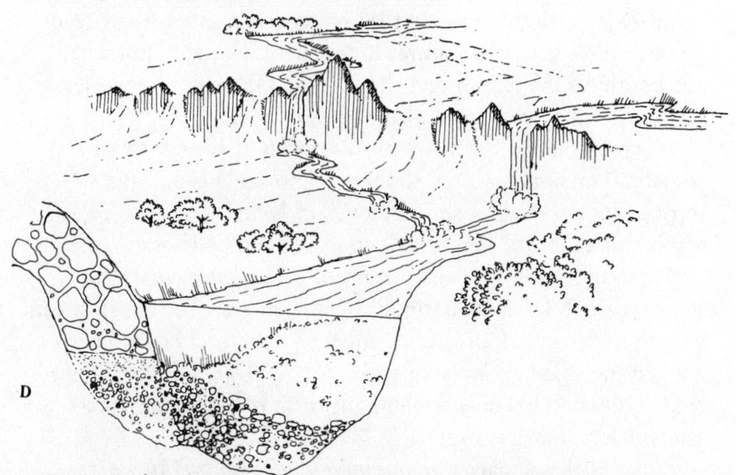

D, the mud has cooled, and a new river has established its way through the greenstone ridge. The original route of the stream behind the ridge is now abandoned. The new stream course crosses the abandoned channel in the background; where it crosses, the stream may steal gold from the older channel. Where the channel lies buried and untouched by the new stream, gold may be locked beneath the ash and volcanic mud.

The cut-out shows an older, buried gravel deposit, an ash bed, a volcanic mud layer, and a new gravel layer. Since the stream now has more water, and therefore more erosive vigor, the new deposit is actually lower than the older one. Part of the older gravel in the bottom of the bed and part of the covering ash have been reworked by the modern stream.

were no bones of dogs, wolves, cats, or other predators, these burials may represent the remains of meals eaten by other, unpreserved creatures of the grasslands.

Although these particular animals did not live at the beginning of this erosive period (they lived in the Pliocene, about 10 million years ago), surely their ancestors lived in the same area 60 million years ago. The little horses eventually gave rise to *Equus,* our horse of today, but the horse vanished from North America with the Ice Age. When the Spanish Conquistadores came to the New World on their fine steeds, no native had seen such an animal.

The courses of the Tertiary rivers that flowed through the mountains and plains to the sea are surprisingly well known, in comparison with the little that is known about other, more readily observable geologic features in the Sierra Nevada. But when one considers the reason that they are well known, it seems less surprising: they contained a great deal of gold.

It is remarkable that a record so tenuous as the course of a stream 50 million years old should be knowable at all. But through long slow days and sudden cataclysms, nature saved enough remnants of the streams to mark their courses.

Thousands of miners and dozens of geologists have toiled many years to trace the antique stream courses. The entire stream bed is not preserved, of course. Most of the gravel beds that choked the channels are long gone – erased by the forces of erosion. Only patches now remain, but from these patches, fossil landscapes can be resurrected.

Some of the patches were preserved beneath lava flows (see Chapter 8); others under volcanic ash falls (see figure 24); some were buried quickly under other sediments. The ash that fell

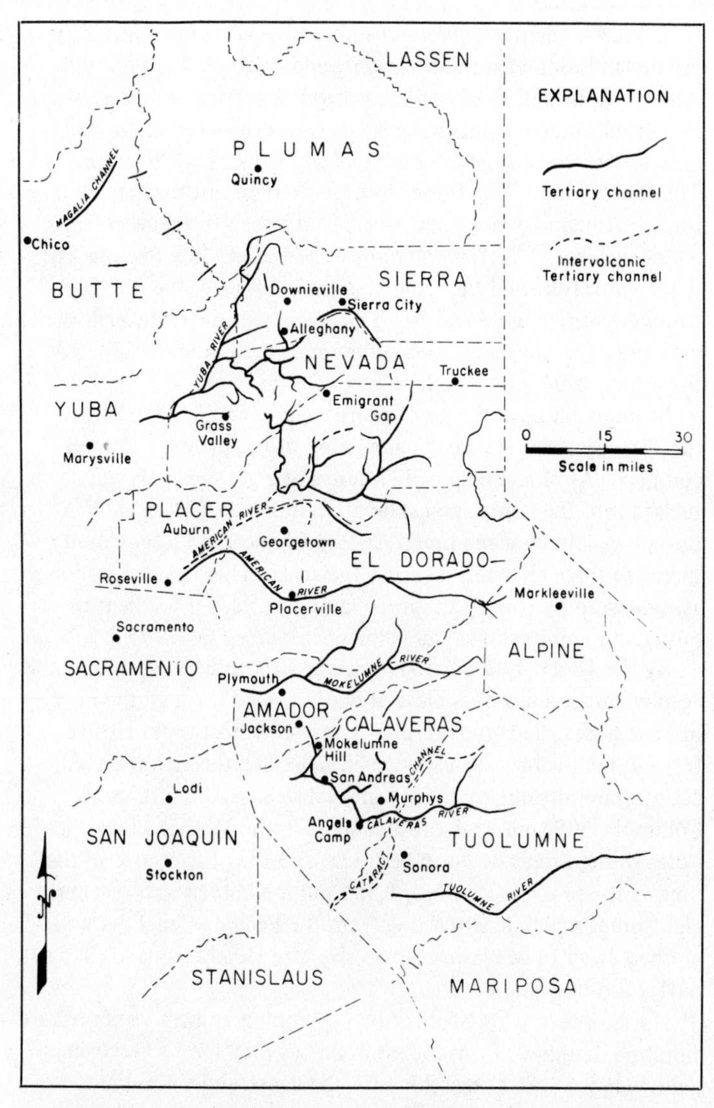

Fig. 25. Sierran rivers of Tertiary time. The oldest channels ("Tertiary channel") are those of 65 to 30 million years ago; the "inter-volcanic channels" are younger, dating from approximately 30 to 3 million years ago (also Tertiary).

from those violent volcanoes of millions of years ago lay lightly on the land in the foothills, blowing and drifting. Where it fell into water, it sank slowly to the bottom, covering and preserving the gravel in the stream channel. In places, it dammed the Tertiary rivers, forming lakes with layers of ash on their bottoms. The tougher, later lava flows, too, filled some of the channels, forcing streams to make new roads to the sea. In this way, a whole new set of Tertiary streams was created. In a few places, the streams followed the same courses; elsewhere, the old channels were buried, gold and all. The new streams cut through the old ones here and there, robbing them of their gold, to add it to the gravel of their own beds.

In some places in the gold country where much of the story of the Tertiary rivers has been deciphered, it has proved to be very complex. At Mokelumne Hill, a series of eight channels was incised into the landscape in the 50 million years from early Eocene to late Pliocene time. At Last Chance, in Placer County, there are three channels; because the streams here gained in power as time went by (owing to more water or a steeper gradient or both), the youngest channel is the deepest (see figure 25).

By the 1860s, hydraulic mining was well underway, ripping the aged channels apart by a blast of water forced through the nozzles of huge hoses. This type of mining was the most rapid erosive force in the history of the Sierra Nevada. All through the gold country are pits dug by these water jets (see figure 26). At Columbia, Volcano, and at other spots in the Mother Lode, the hoses washed right down to bedrock, exposing the course of the long-vanished river. All told, 1,555 million cubic yards of debris – eight times as much as was dug from the Panama Canal – was sloshed away to be washed down rivers to flood farmlands and silt navigation channels.

During the heyday of this firehose mining, miners took several hundred diamonds from the same sluices they used to recover gold. They were not searching for the gems, and since diamonds were not easy to see among pebbles of rock and quartz that looked very much like them, no doubt many more were crushed in the mining process or washed down the rivers.

Fig. 26. How hydraulic mining was done. The sketch is not typical, because most hydraulic monitors (the "nozzle" being directed by the miner) were far too large for one man to control by hand, without a weighted box for leverage. Water workings are shown at the top of the cut; man in bottom is working the sluice box.

In 1884, hydraulic mining was virtually halted by a court injunction that became one of the first legal milestones in the battle to protect our environment. Thereafter, the court ruled, hydraulic miners could mine only if they contained their waste behind suitable dams so as not to clog the navigable waters. Although this meant only the mines on streams tributary to the San Joaquin and Sacramento Rivers (the only two California rivers classed as navigable), few mines survived the ruling. It was too "expensive" to take care, and hydraulic mining essentially ceased. Some mines continued to operate behind dams, but they were small – in no way comparable to the giants of the sixties.

The map on p. 95 shows the courses of the old rivers as they have been pieced together. No doubt there are venerable gravel beds still hidden in the gold country that contain rich "virgin" placers. It is also possible that, if one could trace accurately the shores of the old Eocene sea, he might find deposits of fine gold mixed in with the muds of the deltas of fossil rivers. These are challenges to today's gold miners.

Table 5

Sedimentary Rocks Commonly Found in the Sierra Nevada

Kind of Rock	How to recognize it	Where to see a good example
Conglomerate	Rock is made up of grains 2 mm or more in diameter, together with coarser fragments If fragments are angular, rather than rounded, rock is "breccia," a word that can be used also for volcanic rocks made up of angular fragments	"Auriferous (gold-bearing) gravel;" most is not consolidated enough to be called conglomerate. Some cemented gravel in lower part of Tertiary channels is conglomerate See gravel at Gold Run hydraulic pit, U.S. Interstate 80 (map 4, circle 3); Malakoff State Historic Park (map 2, circle 5); other hydraulic mines Big Chico Creek (map 1, circle 2)
Sandstone	Rock is made up of distinguishable grains, usually somewhat rounded, cemented together; grains can be from 1/16 to 2 mm in diameter Most Sierran sandstone has tiny grains	Folsom (map 4, circle 10) Buildings: State buildings at Ione (map 7, circle 3); Tullock mill, Knights Ferry (map 7, circle 24)

Kind of Rock	How to recognize it	Where to see a good example
Shale	Rock is made up of very fine grains, less than 1/16 mm in diameter Grains cannot be distinguished by naked eye; probably not with hand lens	Dry Creek, west base of Oroville Table Mountain (8 miles north of Oroville) (map 1, circle 4)
Clay	Plastic, pliable when wet Commonly red, white, or gray	Clay pits at Ione (map 7, circle 2), Lincoln (map 4, circle 7) Sierran clay is to be seen mostly in lower western foothills
Lignite	Soft, punky; wood structure visible in some specimens Lignite is soft coal (harder coal is classed as a metamorphic rock)	Buena Vista lignite mine (map 7, circle 16) Mines at Ione (map 7, circle 3) and Carbondale (map 7, circle 1)
Tuff	Since tuff is deposited in layers, it is sedimentary; its source is volcanic	See under volcanic ash and tuff, pp. 122-123
Till	Jumbled mass of clay, sand, and boulders Distinguished from *lahar* by presence of large numbers of nonvolcanic boulders	See under glacial features, table 11.

Chapter 8

DAYS OF FIRE

Violence – fire, ice, and earthquake – are the hallmarks of the past 30 million years in the Sierra Nevada. The record written in the rocks shows it to be a time of volcanic explosions, of creeping molten lava, of roaring volcanic mud flows – a time that transfigured the landscape, revised the courses of rivers, and mantled the mountains with new rock. The long, slow movements of the earth and the wrenching paroxysms of sudden earthquakes lifted the mountains to new heights, even greater than today.

Looking backward through time it seems as though one volcanic episode followed another in rapid succession, and that the Sierra Nevada must have resembled a red hot furnace.

As far as we can tell now, that was not the case. It merely appears to be so because we have records only of "happenings" and little or no record of the long quiet intervals. In our own time, for example, it took but a few months for the volcano, Parícutin, to rise from a level cornfield. Surtsey, in Iceland, was born and grew in the sea while we watched on television; Vesuvius, Krakatoa, Mt. Pelée, and other volcanoes have exploded so suddenly that there was no escape for life at their feet. The record of such eruptions may be thousands of feet of lava solidified upon the landscape, new islands, or ash heaps higher than skyscrapers. These are impressive records, yet they may be the work of only hours or days. Our vision of the past is compressed, like a time-lapse film, by seeing only such sudden episodes.

But if we consider also the quiet intervals in the past thirty million years, the times were more violent, proportionately, than recent times in today's Cascade Range. Within the last century and a half, Lassen volcano and nearby Cinder Cone have erupted; within the past 500 years, other volcanoes in the Cascades have spewed lava and cinders. This is geologically very active, yet we do not think of ourselves as living in the heart of volcanic danger.

Not all volcanoes are equally dangerous. Basaltic ones (see

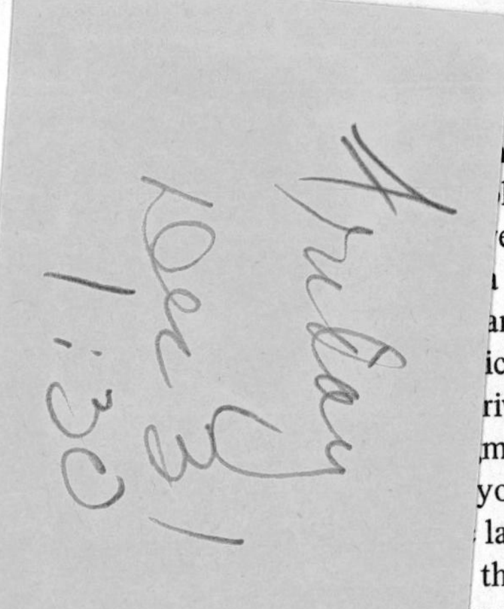

ıt "basic" lava (lava that has a
ı), usually erupt quietly. More
ılitic and andesitic lava (the
'e more silica).
ı – basalt, andesite, and
ıracteristic style of eruption,
ic landscape.
rived from rhyolitic rock are
ments, that poured from fis-
yolitic domes and collapsed
landscapes were the first to be
the Tertiary volcanic episode,
een among the newest to be

ast, tend to be rough and hum-
ows as well as ash falls and the
desite dominates the volcanic

Basalt is the quietest, and from the standpoint of an onlooker, the safest of eruptions. Its landscape features are low domes, lava plains, and gently sloping, gradually built volcanoes. Hawaiian volcanoes, which entertain visitors, are of the basaltic type. Two large basaltic plains cover much of the northern Sierra. Both the older plain, of which Oroville Table Mountain is a part, and the younger plain, in the extreme north, had their origin outside the Sierra Nevada proper. Most other basaltic landscapes in the mountains are smaller, consisting of small "table mountains," cones, columns, or rough-surfaced flows. The most conspicuous of these recent basaltic features have been added to the landscape within the past million years.

The initial phase of the Sierran Tertiary volcanic episode was ushered in by rhyolitic explosions. What happens in a volcanic explosion is that lava in the volcano is torn into fragments of all sizes: dust, sand, blocks, and "bombs." The explosion results from the sudden release of gas (largely steam) from the liquid lava, usually in the throat of the volcano, where the lava is under lessening pressure as it rises toward the land surface. The finest material, called "ash," is blown far and wide by the wind; coarser material is left near the volcano, where it gradually builds the cone higher and higher. Despite their name, volcanic ashes are not the remains of objects burnt by fire, but are minute rock fragments, discernible as rock under a microscope. Some of the material is so fine it can travel thousands of miles: volcanic ash from the explosion of Krakatoa in Southeast Asia, was blown seven times around the world, changing the color of sunsets for more than two years. The sound of the explosion was heard for 3000 miles.

The rhyolitic explosions of the early Tertiary left ash scattered through much of the northern Sierra (see figure 27). Beds of ash as much as 450 feet thick may be seen in the gold country; rhyolite near the mountain crest is in beds as much as 1200 feet thick. Most of the ash (called "tuff" when consolidated) and rhyolite lies north of Yosemite National Park; whether the ash was never scattered beyond, or was so easily removed as to leave no trace, is not known. Since no vestiges remain, likely little of it settled there.

Where the volcanoes were that produced this vast amount of ash is not entirely clear, but it is probable that they stood somewhere along the present Sierran crest and to the northeast. The deposits become less "ashy" to the northeast and are more consolidated, their fragments welded together, indicating that they were hot when they fell and were not cooled by a long trip in the air or in a mountain stream. Miners knew the tuff very well; they burrowed through it to get to the gold-bearing gravel beneath (see drawings on pp. 92-93). In the tunnels that they dug, the tuff was light gray; the same rock, exposed to air, turned to muted shades of cream, violet, yellow, and pink. The rock is very soft, and can be cut with a handsaw and chisel. Many old buildings in the Mother Lode were and are built of it; after it is sawed and dressed, the blocks can be laid tightly with a minimum of mortar. Exposure to the air not only alters their color, but also forms a hard crust that protects them from rapid damage by wind and weather.

Unless they are deliberately destroyed or vandalized, structures built of tuff should last a long time. Ancient Rome was built

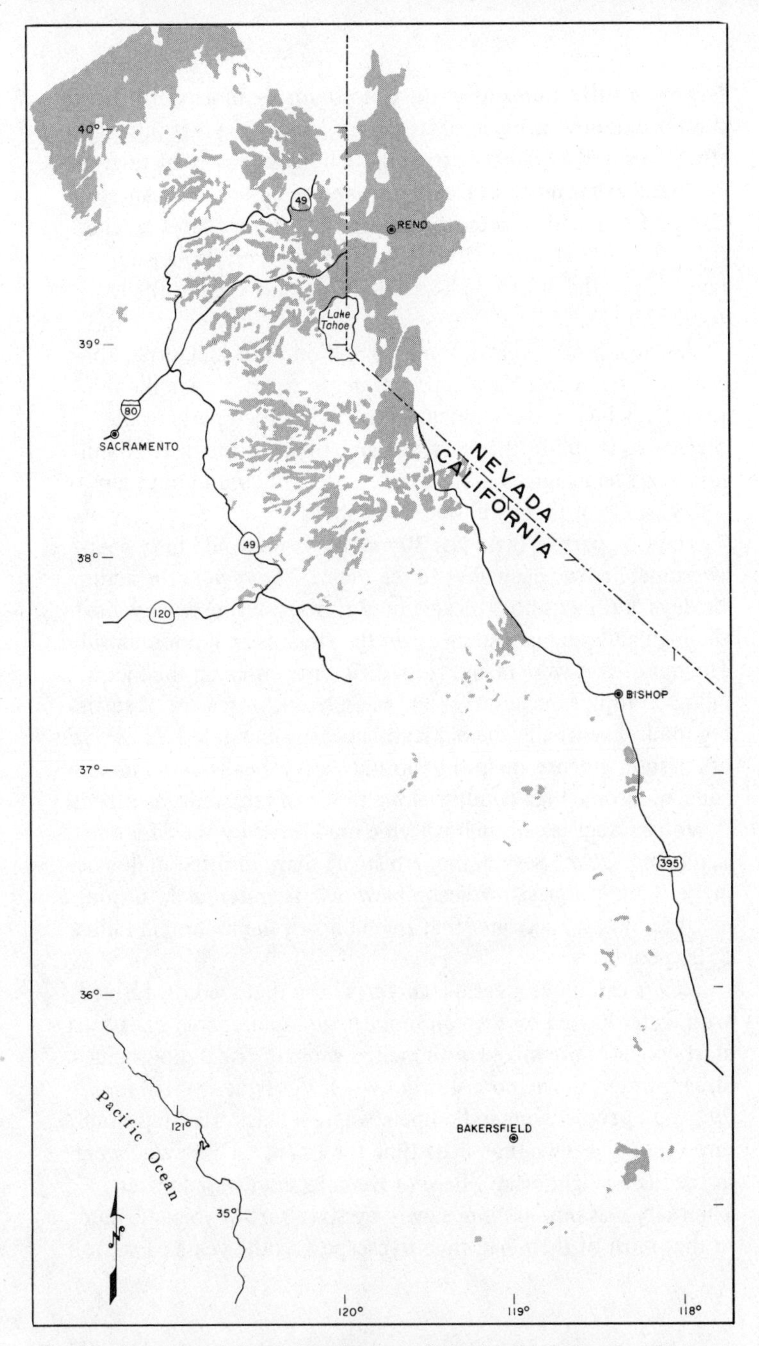

Fig. 27. Outcrops of Cenozoic volcanic rocks in the Sierra Nevada. These are what remain today of the volcanic ash falls, rhyolite flows, and volcanic mud that covered the Sierra from 40 to 3 million years ago.

largely of tuff. Remains of the Colosseum are blocks of tuff, and the Forum, now stripped of its marble facing, is a tuff ruin. Old structures in the gold country, built wholly or partly of tuff, can be found as far north as Dutch Flat and as far south as Angels Camp. Cornerstones, keystones, and lintels show dates ranging from the early eighteen fifties to the nineties; craftsmanship ranges from the most rough-and-tumble stone walls to finely tooled blocks.

As the ash falls waned, some 20 million years ago, other volcanoes in the Sierra Nevada proper began to erupt. Unlike the rhyolite, which took its origin from unidentified parts of the Sierran crest and farther east, volcanic rocks of the next 20 million years had sources that can be traced to Sierran mountains of today, some of them still warm volcanoes.

The first part of these last 20 million years would have been awesome, had we been alive to see them. Those were the andesitic days, a time of hot volcanic mud flows, pouring from vents in the highlands and cascading down the streams and mountainsides. The mud flows were many, from different centers in the Sierra, piling on top of one another and merging to form a sea of steaming mud. Eventually the old landscape was inundated, only ridges of resistant greenstone in the foothills, a few peaks in the middle zone, and some high country along the crest remaining as islands.

Volcanologists call such volcanic mud flows by their Indonesian name, *lahar*. Several modern *lahars* have resulted in disaster: in 1919, an Indonesian volcano blew out its crater lake, forming a mixture of lava and water that roared down the mountain killing 5500 people.

Lahars can move great distances. Those that are hot, formed with water heated by the volcano, can go rapidly; cold ones, that start cooler or are mixed with melted snow or frigid mountain stream water, move more slowly. When Vesuvius erupted in 79 A.D., people living in Pompeii, where a black, flashing cloud enveloped the town, had little time for escape. Those who were in the little neighboring village of Herculaneum were luckier. Their city was buried more slowly by sixty feet of volcanic mud, so that most of them had time to escape. As the years passed, all memory of both cities vanished until they were rediscovered centuries later.

There was no doubt that the mud flow that engulfed Herculaneum came from Vesuvius. We are almost as certain concerning the origin of the lava that formed the Sierran mud flows. In the area around Relief Peak and Mt. Emma, many dikes and plugs of similar composition remain. These and The Dardanelles, Castle Peak, Disaster Peak, Mt. Lola, Mt. Lincoln, Tinker Knob, Squaw Peak, Mt. Mildred, Twin Peaks, Mt. Ellis and others were no doubt vents for the many flows.

In the higher country, or near the volcanic vents, the lava, while still hot and liquid, was mixed with water, giving it enough mobility to travel downhill as a slurpy mud. Rain, snow, lakes, and rivers were the source of water. Sometimes the lava flowed in the stream channel itself, mingling with stream water to make a steaming brown paste. In some places, mud flows dammed the stream channels, forcing them to take other routes to the sea (see drawings, pp. 106-107).

The mud flows poured down the mountain slopes as jumbled masses of large rocks and small, of boulders of granite and metamorphic rocks, of torn and shattered trees, mixing and churning, coming to rest when they had stiffened too much to move farther or had reached the bottom of the slope. Landscapes that developed on the mud flows have a chaotic aspect. Travelers through Sierran passes drive through andesitic terrain, sparsely populated with bushes and digger pine, for mile on mile. One particularly striking outcrop is at Carson Spur on State Highway 88. Here the mud flow remains are perched above a branch of the American River, which runs hundreds of feet below through granite. Ages of weathering have worn the face of the cliff back, carving the volcanic rock into battlements and towers that resemble a derelict fortress.

The andesitic mass is sometimes called "andesitic breccia," a term indicating that the rock has been broken apart. Why it should be broken is not entirely clear: perhaps the over-and-over downhill tumble shattered it; perhaps internal pressure that built up within it caused it to break apart; or perhaps both.

Fig. 28. Stages in the creation of Tuolumne (Stanislaus) Table Mountain. *A* shows the Tertiary Stanislaus river flowing through moderately rolling hills, flanked by subtropical vegetation. In *B,* steaming lava has followed the easiest course downhill – the bed of the river. It has filled, but not overflowed, the river valley. The third view, *C,* shows the Tertiary Stanislaus river course as it appears today near Knight's Ferry. The lava flow, being much harder than the surrounding hills, still marks the course of the ancient river. The softer rock of the enclosing hills has been eroded away, leaving a high, sinuous ridge where once there was a river valley.

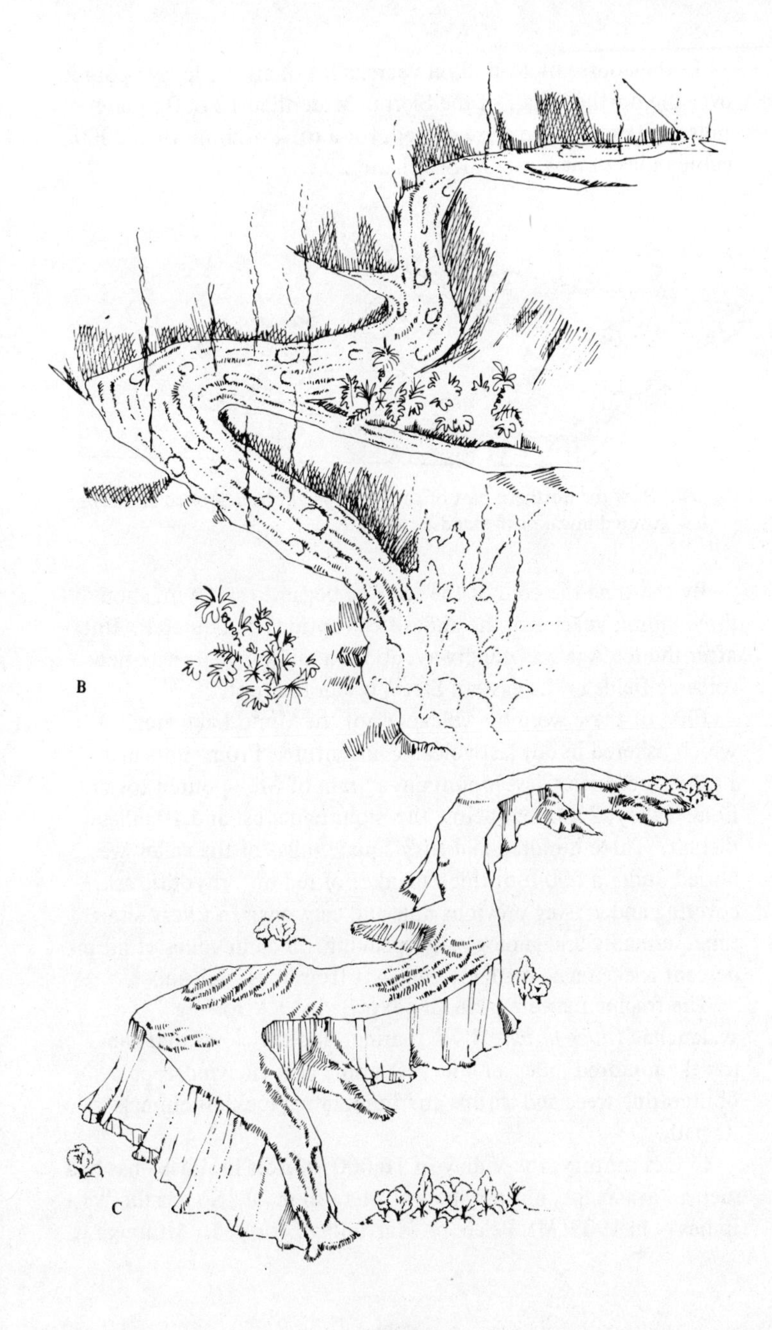
B
C

In the course of 15 million years, a lot of andesitic lava poured over the northern half of the Sierra. More than 12,000 square miles of land surface was covered, for a total of more than 2000 cubic miles of new rock! (See figure 29.)

Fig. 29. How the northern part of the Sierra might have looked after *lahars* had covered much of the landscape.

By the time the cold of the Ice Age began to settle in, about three milion years ago, the andesitic eruptions had quieted. But after the Ice Age was underway, about a million years ago, new volcanic fields in the eastern Sierra began to erupt.

First of these were the volcanoes of the Mono Lake area, which ushered in our last volcanic adventure. From vents near 11,000-foot-high Glass Mountain, a "rain of fire" poured toward Long Valley, 2100 feet below the mountain crest and 10 miles distant. Three hundred and fifty square miles of the valley was buried under a 500-foot-thick blanket of red-hot, rhyolitic ash, covering and erasing previous hills and canyons. In a very short time, humanly and geologically speaking, 35 cubic miles of incandescent rock fragments had been shot from these volcanoes.

The fragments were probably expelled as a "glowing avalanche" (*nuée ardente*) – a roaring, flashing cloud that can travel a hundred miles an hour, sweeping over the landscape, obliterating trees and shrubs and immolating luckless animals in its path.

In this century, the Valley of 10,000 Smokes in Alaska has had such an avalanche, but no one was there to see it. So has the West Indies – in 1902, Mt. Peleé, on Martinique, erupted. Although it had been giving premonitory rumblings, with ashes falling in the streets and horses suffocating in their tracks, the signs were not heeded by many for various personal and political reasons. Suddenly, at 7:50 A.M. on May 8, the volcano exploded with four deafening blasts. A *nuée ardente* sped down upon the city of St. Pierre, leaving but two persons alive – one a murderer awaiting execution, protected in his dungeon from the death that came to others. All of the houses in St. Pierre were destroyed. Thirty thousand people lay dead, within minutes. The dull red cloud, pierced with brighter red streaks as larger incandescent rocks were tossed about in its billows, raced to the sea where it sent up clouds of steam two miles high.

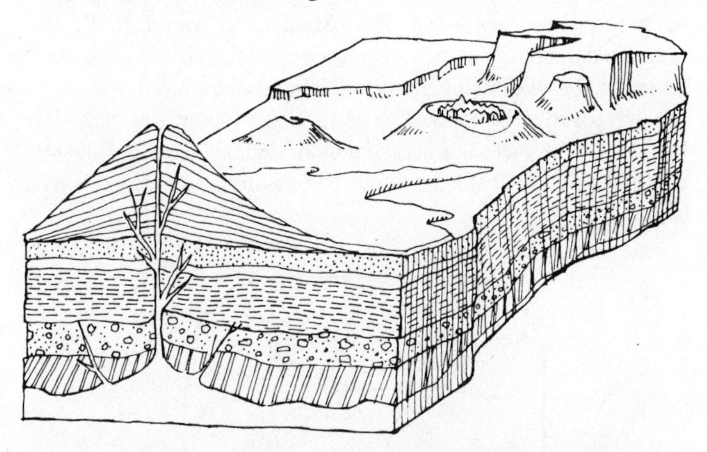

Fig. 30. Block diagram showing volcanic landscape features. Magma, or molten rock, rises to the surface and into the overlying sedimentary rock along planes of weakness. Within the earth, it forms dikes and related igneous bodies; that part that pours out upon the surface creates various volcanic land forms.

Among the land forms are volcanoes (left). From a vent on the opposite side of the volcano, a lava flow has spread out over the land.

Lava plateaus and lava mesas are in the background, while ash cones, cinder cones, and volcanic domes are in the middle ground.

All have their source in the magma chamber beneath. The magma chamber itself may eventually solidify completely. When unroofed of its overlying volcanic and sedimentary rock, the solidified magma chamber probably would be a coarse-grained igneous rock, possibly diorite or granite.

Truly, in such an eruption, the volcano looks as if it were on fire. But it is not burning in the sense that it burns objects "up," except for trees that the cascading lava may encounter. The red fiery color is even more frightening than fire: it is the color of rock heated to incandescence – to a temperature not even approached in furnaces or fireplaces where wood or coal is being burned.

In the Sierra Nevada, the temperature of the center of the ash layer when it was deposited in Long Valley must have been at least 650°C. (1200°F.). It may actually have been twice that hot. The ash settled to perhaps half its original thickness as it welded and compacted. The tiny glass fragments of which it was made fused together as it cooled. Similar tuff from Hawaii, when magnified many thousands of times, resembles childrens' jacks (see figure 31); no doubt the Sierra tuff did, also, before it was crushed together by its own weight. Today, some 700,000 years later, after the sheet has cooled and eroded, an air view still shows a remarkably even table land, cut through by a few now-dry rivers (figure 32).

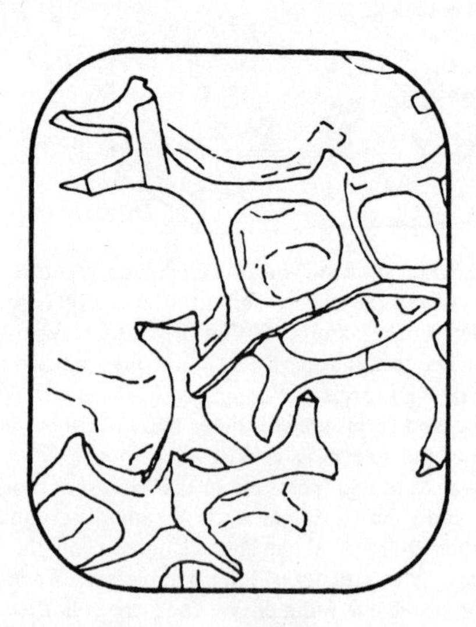

Fig. 31. Fragments of volcanic ash, magnified many thousands of times.

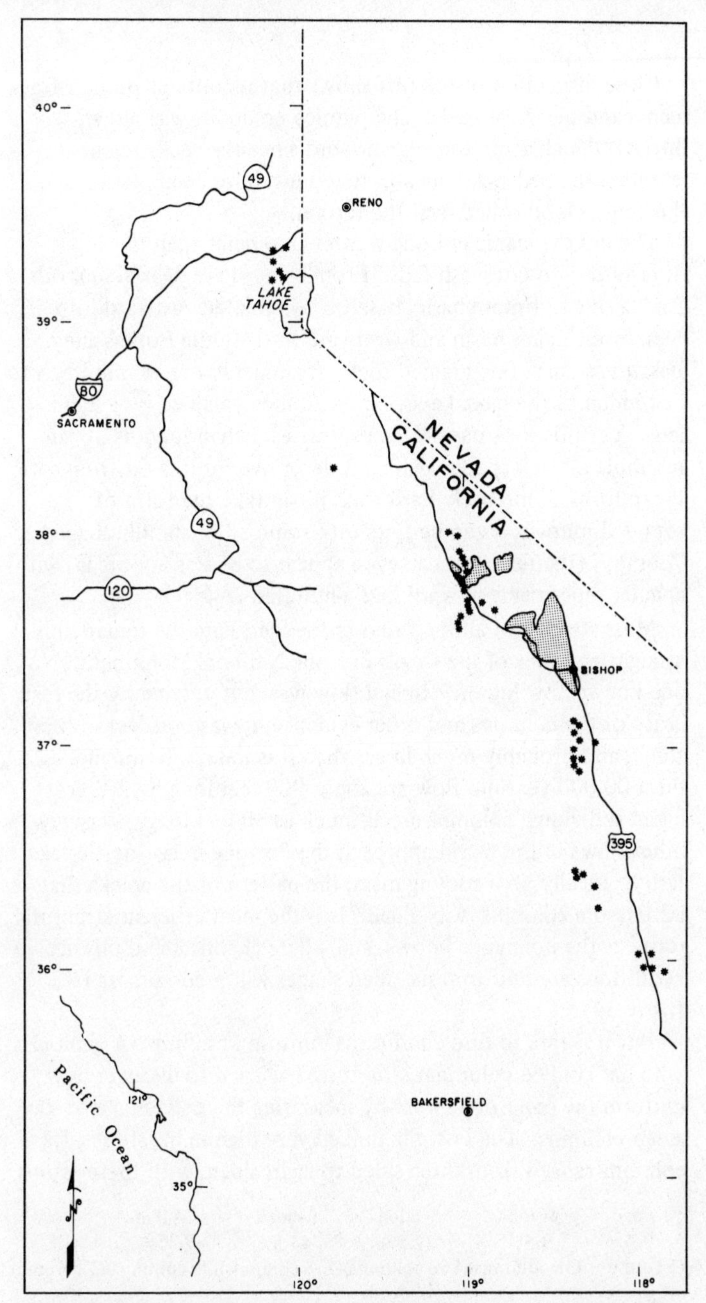

Fig. 32. Extent of the remains of the *nueé ardente* of 700,000 years ago in eastern California (stippled; these are outcrops of the Bishop Tuff). Asterisks mark volcanic centers that have been active in the past 3 million years.

Close inspection of the tuff shows that it contains pumice fragments and pieces of rock. The pumice no doubt was blown through the air from the volcano, but since the rock fragments resemble the bedrock beneath, they must have been picked up as the dense cloud rolled over the terrain.

The next volcanic episode was much quieter than the devastating rhyolitic ash falls. From somewhere near Mammoth Pass, a sheet of more basic, basaltic lava poured eastward into Mammoth Lakes basin and westward into Middle Fork Valley. Basalt is a dark, fine-grained rock. Its minerals are so minute as to be hidden to the naked eye – even to the eye aided by a hand lens. Yet this rock has protean surprises: when it cools, it can resemble a black sponge, which is deceptive for it is far from soft; it can form an intensely hard wall; it can take the form of rounded pillows; it can be spun into rounded or spindle-shaped "bombs" (figure 33); it can even appear to be a stony organ, with angular pipes rising upward half-a-hundred feet.

Most striking of all the flows in the Sierra are the remarkably straight columns of the Devils Postpile National Monument. No one knows how big the original flow was, but judging by the regularity of the columns and other evidence, it was 600 feet or more thick, and probably much larger than it is today. Remnants of this 600,000-year-old flow are about 900 feet long by 200 feet high. Individual columns are as much as 60 feet long. Very few other flows in the world approach the Postpile in geometric regularity. Ideally, in a cooling mass, the pattern of the cracks that initiate the columns is six-sided. It is the most efficient structural form, as the honeybee knows, and where chemical and physical conditions are uniform, six-sided shapes will predominate (see figure 34).

But it is rare to find conditions uniform in nature. A geologist who has studied columnar structures has tried to discover how uniform the conditions were by measuring the percentage of six-sided columns. The Postpile ranked very high in his study. Its columns ranged from three sided to eight sided, with these ratios:

3-sided	4-sided	5-sided	6-sided	7-sided	8-sided
½%	9.5%	37.5%	44.5%	9.25%	¼%
(2 columns)	(38 columns)	(150 columns)	(178 columns)	(37 columns)	(1 column)

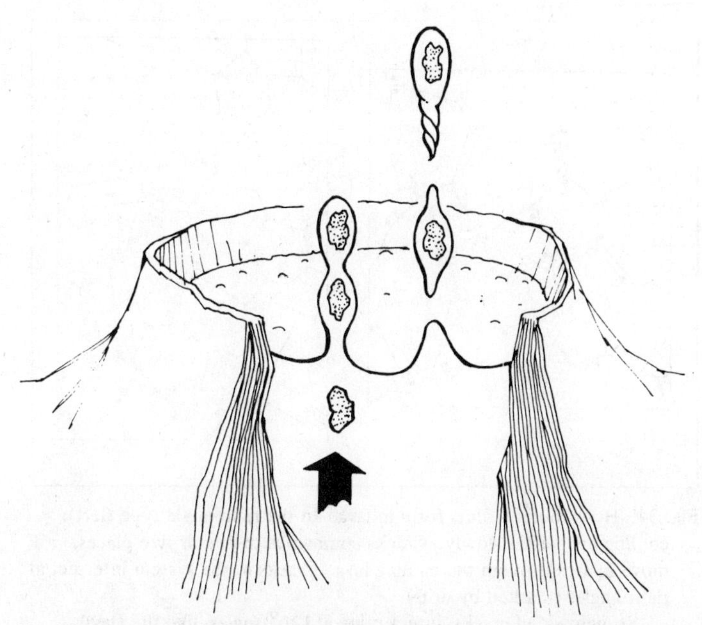

Fig. 33. How volcanic bombs are formed. Ribbons of lava, some surrounding cold rock fragments, are blown upward from the lava lake in the volcano. They may be sent spinning 4000 feet into the air at speeds of as much as 375 miles per hour. Not to scale. Bombs range in size from a few inches to several feet in diameter; volcanoes may be miles across.

Another person counting found 55 percent six-sided columns.

Compare these figures with statistics from other well-known columns: Giant's Causeway, Northern Ireland, six-sided columns, 51 percent; Devil's Tower, Wyoming, 32.5 percent; Craters of the Moon, Idaho, 16 percent.

The angles at which the sides of the columns meet also give clues to their cooling history. In piles with straight sides, like the Postpile, the angles tend to be close to 120°; piles with more 90° angles tend to have curved faces.

Crack (joint) systems like the Postpile probably developed more or less instantaneously. Conditions were so uniform throughout the cooling mass that all parts of it reached crack-forming stress at the same time. The whole system then cracked more or less at once. This contrasts with the right-angled system,

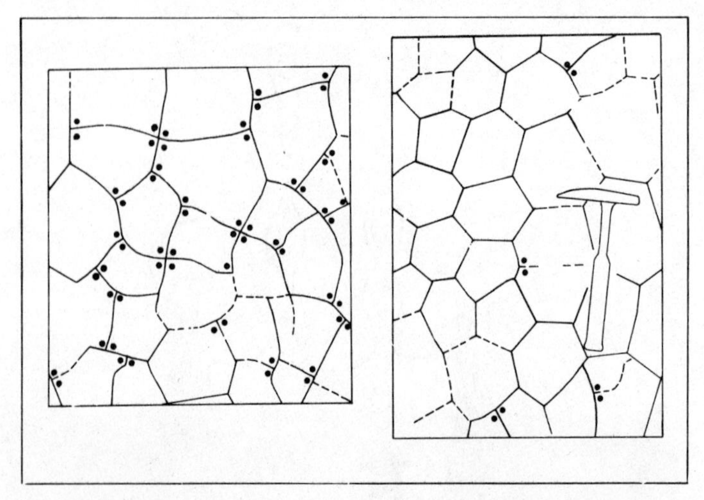

Fig. 34. How cooling joints form in lava. In the right-angle type (left), cooling progresses slowly. Cracks commence in one or two places, move in arcs through the cooling lava. Cracks in the system intersect at right angles (marked by dots).

Straight-sided cracks, intersecting at 120° angles, like the Devils Postpile (right), are more rare than curved ones. They form when lava cools more or less at once. Since the top and bottom of a flow are in contact with cooler air or colder rock, the pattern is less regular there than in the center of the flow. Geologic pick gives scale.

where a crack will make its way, then stop until conditions change; a branch will form, then another, and so on, slowly.

The piles, as one might expect, form perpendicular to the bottom and top of the lava flow. Probably the part near, but not on, the bottom of the flow was the best place for uniform cracks to start. Where the ground surface was uneven, the piles were bent.

Although "columnar basalt" is a familiar term, and the rock at the Postpile is basalt by every field and mineralogical test, chemical analysis shows it to be more closely related to andesite. It is not possible to tell this by looking at the rock, however.

From below, the Postpile looks like a gigantic organ; from above, glaciers have worn and polished the mosaic to a parquet-like surface.

Mammoth Mountain, east of the Postpile, was an operating volcano about the time or shortly after the piles were formed. It commenced about 400,000 years ago (see figure 35), a big plug dome that pushed its way up as a sticky mass. New flows came up through the old ones to pour out over the sides, forming thick, stubby glass lumps. At least one flow was fluid enough to run

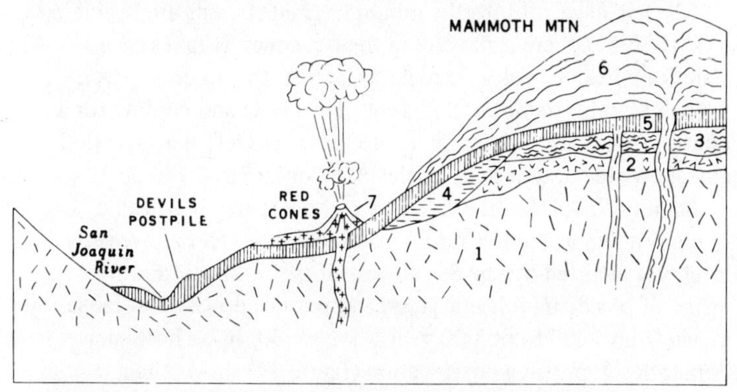

Fig. 35. The sequence of volcanic rocks and events in the Mammoth Mountain – Devils Postpile area. *1* marks the older metamorphic rocks and the granitic core of the Sierra Nevada. Above it is *2*, an andesitic lava flow accompanied by volcanic ash and cinders, about 3 million years old. This flow is especially visible at Deadman Pass. It was followed and overlain by *3*, a lava flow originating near Two Teats. *4* marks the remains of a *nueé ardente* – glowing avalanche – at Reds Meadow (Devils Postpile National Monument). Deposited less than a million years ago, this ash deposit filled the canyon of the Middle Fork of the San Joaquin to a depth of at least 1,000 feet. Much of it has been eroded away.

Next came the Devils Postpile flow itself *(5)*, issuing from the area near Mammoth Pass. When it was first deposited, about 600,000 years ago, the flow was 600 feet or more thick; erosion by streams and glaciers has considerably lessened that. About 400,000 years ago, after the formation of the Postpile, Mammoth Mountain began to grow by piling up thick, glassy volcanic flows, *6*.

Glaciers formed on the mountain during the last part of the Ice Age, but volcanic explosions have blown holes in the side of the mountain since the glaciers melted. Steam from fumaroles on the side indicate that the volcano is not yet extinct. The last major volcanic episode, *7*, resulted in the formation of two basaltic cones, Red Cones, three miles south of the Postpile.

two miles beyond the base of the mountain, but most merely stacked another layer on top of the older ones.

But Mammoth volcano is not simply layer after layer of lava, like icing on a cake. Several times explosions blew holes in it, and at other times, ash and pumice poured out of vents. It reached its greatest height about 370,000 years ago.

Glaciers have cut into the mountain, and streams are beginning to slice it, but there is life still in the old cone. It takes a long time for a heap of volcanic rocks as large as this to cool. Even though Mammoth Mountain has been erupting and cooling for a quarter of a million years, there are fumaroles left that occasionally show steam, especially in the cold winter air.

Mammoth volcano is related to a whole string of volcanic domes known as Mono Craters. Standing 9000 feet above sea level, yet dwarfed by the Sierra west of them, the craters are a string of obsidian (volcanic glass) and pumice domes that range in age from 1300 to 60,000 years. Figure 36 shows how such domes develop, and a cross-section (figure 37) shows their structure.

That the whole field is far from dead is demonstrated by the many hot springs up and down the Sierran front. They, like the volcanoes, are more or less in a line. The warmth of their waters means that they are in touch with some subterranean heat source – no doubt a slowly cooling magma body.

When will the next eruption be? No one can now say. When the last eruption in the Sierra Nevada took place is not known for certain; the last event we know of was an underwater eruption in the bottom of Mono Lake in 1890. No one saw it, but steam and sulfurous fumes rose from the lake in puffs, and normally cold water springs spouted boiling water and mud. What did happen there in the depths of the lake? A lava flow? An underwater volcano, twin, perhaps, to black Negit, now an island in the lake? Recent bathymetric surveys show islands and basins on the lake floor; one of them may have been the culprit. Near Negit, hummocks on the bottom do look like submarine lava flows that escaped from Negit or from underlake sources.

It would be interesting for a skin diver to search for new "pillows" on the lake bottom, if the water were not too deep and

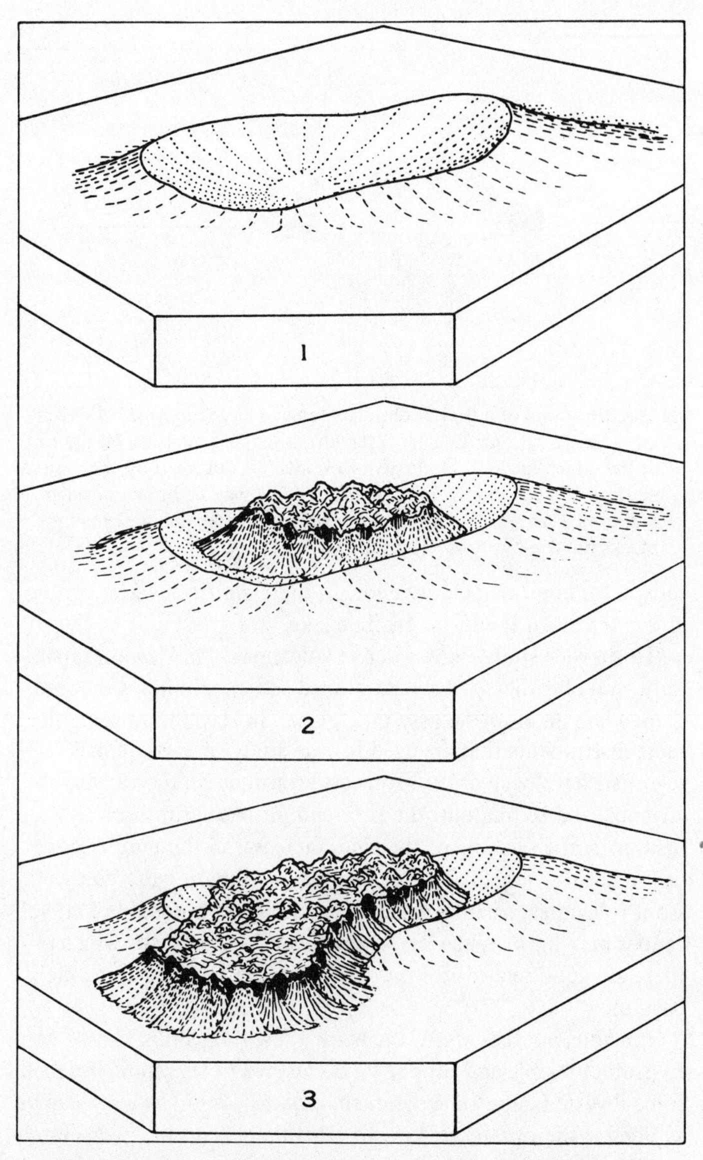

Fig. 36. How an obsidian dome develops. *1*, a low cone has formed by explosions of pumice. *2*, a volcanic dome rises in the center of the cone, pushing upward. *3*, the dome has risen so high that part of it has pushed over the crater rim as a short, thick flow called a *coulée*. Blocks of obsidian dot the *coulée* and flow.

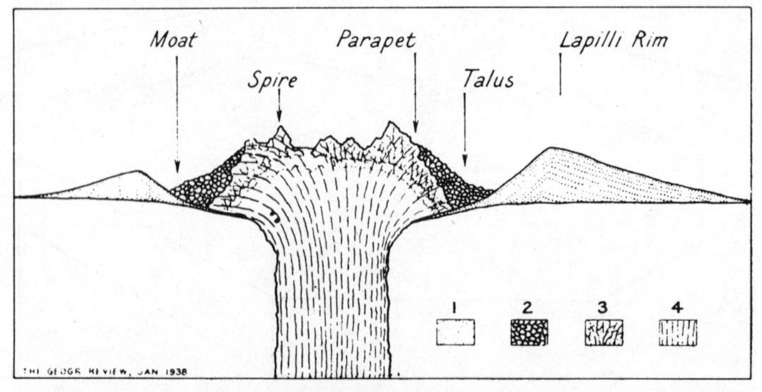

Fig. 37. Structure of a typical obsidian dome. Outer rim, formed of fine ash or small volcanic fragments (lapilli), is marked by dots *(1)*; the core of the dome, made of black, glassy obsidian, is indicated by the pattern marked *4*. The more frothy top of the dome may be brown obsidian, indicated as pattern *3*. As the dome rises, and later as it weathers, blocks break off to form heaps of rubble (talus), *2*.

caustic for comfort. Oddly enough, there are fresh water springs in the depths of the bitter alkaline lake.

Those who study known active volcanoes – in Hawaii, Japan, Italy, and elsewhere – have developed some techniques that tell them when an eruption may take place. In Hawaii, by using the same instruments that are used for the study of earthquakes, scientists have been able to forecast eruptions six months in advance, and to pinpoint the time and place of eruption. By measuring the swelling of the land, increases in the temperature of hot springs, and changes in the composition of gases escaping from volcanoes, we may get some warning. It is possible that volcanoes may have regular cycles, but even Vesuvius, for which there are good records for more than 1000 years, shows no clear pattern.

If a volcano does erupt, there are a few things that can be done to protect people and property. Some lava flows can be bombed, some diverted, and some dammed. *Lahars* – mud flows – can be avoided if their paths are known. But there is nothing, absolutely nothing, that can yet be done in a *nuée ardente,* except to heed early warning signs of eruption and flee.

If an eruption of the magnitude of the 1912 eruption of

Katmai, Alaska, were to take place in populous California, the disaster would be incalculable. In historical perspective, though, our astonishing luck has, to now, kept us from counting volcanic eruptions as major disasters. They are far, far behind wars, automobile accidents, hurricanes, floods, and earthquakes.

Table 6

Volcanic Igneous Rocks Commonly Found in the Sierra Nevada

Kind of Rock	How to recognize it	Where to see a good example
Rhyolite (felsite)	Light colored Fine grained; most minerals too small to be distinguished even with a hand lens; a few larger fragments of quartz, feldspar, or pumice may be visible Rock may be banded Generally very hard May ring when struck by hammer	Wilson Butte, U.S. Highway 395 north of Deadman Summit (rhyolite glass)(map 9, circle 31) "Petroglyph Loop Trip," near Bishop (rhyolite tuff)(map 12, circle 5); see also "volcanic ash" Silver Peak, Ebbett's Pass (map 6, circle 8) Highland Peak, Ebbett's Pass (map 6, circle 9) Lookout Mountain (map 9, circle 34)
Andesite	Medium gray Most minerals too small to be distinguished even with a hand lens; a few crystals of dark minerals may be visible Chip or rock held to light is translucent on edge	Tuolumne (Stanislaus) Table Mountain ("latite")(map 8, circle 8) Highway 88, near Carson Spur (map 5B, circle 1) Devils Postpile (map 11, circle 2) See also *"lahar"* and "dome," table 7

Kind of Rock	How to recognize it	Where to see a good example
Basalt	Dark colored; resembles andesite, but is usually more nearly black Chip of rock held to light is opaque May form columns	Honey Lake (on index map) Oroville Table Mountain (map 1, circle 5) Golden Trout Creek (map 14, circle 16) Sawmill Creek (map 12, circle 16) See also "lava flow," table 7
Obsidian	Glassy Rounded fragments chip off like glass (conchoidal fracture) Sharp; will cut flesh Generally dark colored	Glass Creek flow (Obsidian Dome), off U.S. 395 near Deadman Summit (map 9, circle 32) Mono Craters (map 9, circle 28)
Pumice	Gray or yellowish gray May have many holes (vesicular) Floats on water (formed as lava froth)	Pumice Flat, Devils Postpile National Monument (map 11, circle 1) Devil's Punchbowl, Mono Lake region, U.S. Highway 395 (map 9, circle 29) Mono Craters (map 9, circle 28) Pumice Butte (map 11, circle 8)
Cinder	Pebble-sized fragment, resembling very small furnace "clinker" Usually red or black (red results from oxidation – rusting – of iron)	See under "cone," table 7
Tuff	Very fine grained; grains not visible under hand lens Usually light colored (Microscope shows tuff to consist of very tiny pieces of volcanic glass or rock or both)	See "rhyolite," this table, and "volcanic ash," table 7

Metamorphic rock of the Sierra Nevada. Although "tombstone rocks," like those above, appear to rise from the ground, quite the reverse is true: they are the now-vertical stubs of rock that once was formed as a layer on the sea floor. These are probably the relics of undersea volcanoes. (Mariposa County foothills.) Layered rock (left), now metamorphosed, gives a better idea of what the metamorphic rocks once were. These are among the oldest in the Sierra (Ordovician, about 450 million years old; Mt. Morrison area). The bottom closeup shows how minerals tend to become elongated during metamorphism, and tend to be oriented in one direction, giving the rock "schistosity." (Shadow Canyon, near Devils Postpile.)

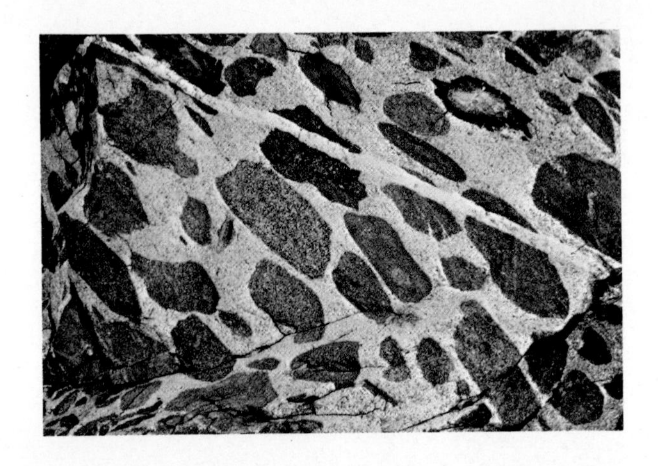

Granitic rock in the Sierra Nevada. Domes like Lembert (below) are far less jointed than the rock of Mt. Whitney (far right). Glaciers and weathering by water have given them an unusual, rounded aspect, unlike the splintery faces of the high peaks to the east. This dome is made of granite porphyry – a rock that resembles granitic rock in texture and composition, but one that also has feldspar crystals as large as several inches in diameter – conspicuously larger than the grain of the rest of the rock.

Above, gray Sierran granitic rock has many inclusions: some are other types of granitic rock, others fragments of metamorphic rock. In places, the granitic rock has a freckled appearance owing to the number of inclusions.

A striking contrast (above) between the sharp, pointed peaks of the High Sierra and the Alabama Hills, at their eastern feet. Moisture from the ocean is carried upward over the bulk of the Sierra. Most of it falls on the western side or near the crest, making the eastern slope into a desert, as the vegetation clearly shows. The Alabama Hills (also below) are formed of Sierran granite, but they have a different aspect. They have not been glaciated; their forms, which are rounded rather than sharp, take their origin from desert weathering.

Preceding page: Yosemite Valley in winter.

The Devils Postpile from near the top. (Devils Postpile National Monument.)

Above, top of Postpile. A glacier has polished this parquetlike floor. Center, lava as it appears in young flows along U.S. Highway 395. Bottom, eroded battlements formed of volcanic agglomerate – solidified volcanic mud flow.

In the photograph in the center of this color section, winter in Yosemite Valley is reminiscent of glacial days. A hanging waterfall, marking the site of a glacier tributary to the trunk glacier that occupied Yosemite Valley, reminds us of the ice that has vanished. Above, little Walker Lake (center of photo) is dammed by a glacial moraine (dark green); behind is the edge of Mono Lake, once filled to overflowing by glacial meltwater. Left, the combination of large and small rocks that makes up glacial till; bottom, a surface of granite, polished by glaciers.

Table 7

Volcanic Features of the Sierra Nevada

Feature	Where to see a good example
Volcano	Mt. Rose, Nevada (map 3, circle 2)
	Mammoth Mountain, near U.S. Highway 395 (map 11, circle 6)
	Negit Island, Mono Lake (map 9, circle 5)
Dome	Templeton Mountain, Monache Mountain ("latite" – andesite) (map 14, circle 8)
	Markleville Peak, Alpine County (andesite dome) (map 6, circle 5)
	Silver Peak, Ebbett's Pass (carved from rhyolite dome) (map 6, circle 8)
	Highland Peak, Ebbett's Pass (rhyolite dome; cinder cone on one side) (map 6, circle 9)
	Glass Mountain, U.S. Highway 395 (obsidian, 1 million years old) (map 9, circle 30)
	Mono Craters (rhyolite tuff rings, domes, and flows, 6400 to 10,000 years old) (map 9, circle 28)
	Panum Crater (dome) (spines, ridges, bombs, pumice, 1300 years old) (map 9, circle 27)
	Wilson Butte, U.S. Highway 395 north of Deadman Summit (rhyolite) (map 9, circle 31)
	Jackson Butte (map 7, circle 8), Golden Gate Hill (map 7, circle 14), Tunnel Peak (map 7, circle 12), McSorley Dome (map 7, circle 11), Hamby Dome (map 7, circle 13), all near Mokelumne Hill (older, eroded andesite domes)
Cone	Red Mountain, Owens Valley (cinder cone 600 feet high) (map 12, circle 15)
	Red Hill, north of Little Lake, U.S. Highway 395 (mined for cinders) (map 14, circle 10)
	Black Point, Mono Lake (basaltic cinders) (map 9, circle 4)
	Red Cones, middle fork of San Joaquin River (map 11, circle 5)
	Lake Tahoe (cinder cone used for sewage disposal) (map 3, circle 3)
	Headwaters of south fork of Kern River (less than 1 million years old) (map 14, circle 7)

Feature	Where to see a good example
	Pumice Butte, near Mono Craters (pumice) (map 11, circle 8)
	Paoha Island, Mono Lake (map 9, circle 6)
Lava flow	Piles at Devils Postpile National Monument (columnar structure; parts of flow show pillow structure where lava flowed into water) (map 11, circle 2)
	Oroville Table Mountain, Butte County (olivine basalt) (map 1, circle 5)
	Tuolumne (Stanislaus) Table Mountain, especially accessible at junction of State Highways 108 and 120 (9 million years old) (map 8, circle 8)
	Sonora Peak (map 6, circle 13), Leavitt Peak (map 6, circle 14) (sources of Tuolumne Table Mountain flows)
	Sawmill Creek, U.S. Highway 395 (spongy appearing lava, bombs) (map 12, circle 16)
	San Joaquin River headwaters, especially near Pincushion Peak (map 11, circle 15) and Saddle Mountain (map 11, circle 16), Kaiser Peak quadrangle (2-4 million years old)
	Mt. McGee, Deadman Pass (2-4 million years old) (map 11, circle 12), Golden Trout Creek, south of Mt. Whitney (map 14, circle 6) (columnar structure)
	San Joaquin Mountain, John Muir trail, near Ritter Range (columnar structure, pumice) (map 9, circle 35)
	Dardanelles, Sonora Pass, State Highway 108 (map 6, circle 11)
	Columns of the Giants, Sonora Pass, State Highway 108 (columnar structure, 150,000 years old) (map 6, circle 12)
	Glass Creek obsidian flow (rhyolite glass; 150-200 feet thick; banding common) (map 9, circle 33)
Lahar (volcanic mud flow)	Two Teats, Mt. Morrison 15-minute quadrangle (about 3 million years old) (map 9, circle 37)
	Carson Spur, State Highway 88 (map 5B, circle 1)
	Thimble Peak, State Highway 88 (map 5B, circle 2)
Volcanic ash and tuff (remnants of *nuée ardente*)	Buena Vista Peak (map 7, circle 16) and Valley Springs Peak (map 7, circle 15) near Valley Springs, Amador County (20-30 million years old)
	Quarry, east of Altaville on Murphys road (20-30 mil-

Feature	Where to see a good example
	lion years old) (map 7, circle 18) Exposures on U.S. 395 near Bishop (map 12, circle 5); along shores of Lake Crowley (map 12, circle 1); on Owens (map 12, circle 2) and Rock (map 12, circle 3) Creeks. (Petroglyphs of "Petroglyph Loop Trip" are carved in 700,000-year-old tuff) Sotcher Lake, Devils Postpile National Monument (map 11, circle 3) Reds Meadow Ranger Station, Devils Postpile National Monument (welded tuff; very hard; sealed together as it cooled) (map 11, circle 4) *Buildings:* Alpine County Courthouse, Markleville (map 6, circle 6) Silver Mountain City Jail, State Highway 4 (remnants only) (map 6, circle 7) Town of Mokelumne Hill (most stone buildings) (map 7, circle 9) Town of Murphys (many stone buildings) (map 8, circle 2) Town of Angels Camp (most stone buildings; good example is Lake's Hotel) (map 7, circle 20) Prince and Garibardi store, Altaville (map 7, circle 19) IOOF Hall, Jackson (map 7, circle 7) Douglas Flat (stone building) (map 8, circle 3)
Explosion pits	Inyo Craters, near Mammoth Mountain, U.S. Highway 395 (500 to 800 years old) (map 9, circle 45) Paoha Island, Mono Lake (map 9, circle 6)
Bomb	Sawmill Creek lava flow, U.S. Highway 395 (map 12, circle 16)

Chapter 9

DAYS OF ICE

In the great, worldwide Ice Age that ended a few thousand years ago, glaciers pushed their way across Canada and much of eastern North America. In the west, they covered the northern parts of Washington, Idaho, and Montana. The Rocky Mountains, the Sierra Nevada, and the Cascades had, by then, become high enough barriers that the sheets did not cross them; but these ranges were also high enough and cold enough to accumulate glaciers of their own.

These "mountain" or "valley" glaciers of the west, although they existed at the same time as the great continental ice sheets to the north and east, were wholly independent of them. Nevertheless, when glaciation reached its maximum, the high Sierra must have been white both winter and summer, for the "firn line" – the line marking permanent ice – covered most of the high country, leaving only ridges and peaks projecting above it (figure 38). These mountain islands, rising from a bath of ice, were frozen, barren deserts. Because of the intense winter cold, the bitter winds whipped the summits more wildly than today, removing snow from such high peaks as Whitney as fast as it fell.

Below, in the canyons to the west, the snow fell thick and deep. "It is hard," wrote John Muir, "to realize the magnitude of

Fig. 38. The firn line (limit of permanent ice) as it was 20,000 years ago in the Sierra Nevada. The area above the firn limit is the mountain ice cap. Not all of the area above this line was covered by ice; isolated, wind-swept peaks of the high crest projected above. Small glaciers also formed in cirques outside that limit. Areas designated as "Pleistocene glacial deposits" are the moraines and associated debris piles left by Ice Age glaciers, still visible in the landscape. Present-day glaciers, marked by dots, are younger than the Pleistocene deposits. Although they hang in cirques carved by Ice Age glaciers, they are the product of the "Little Ice Age" and were at their height only about 200 years ago. There are about 70 small glaciers in the Sierra Nevada; only general areas are shown here.

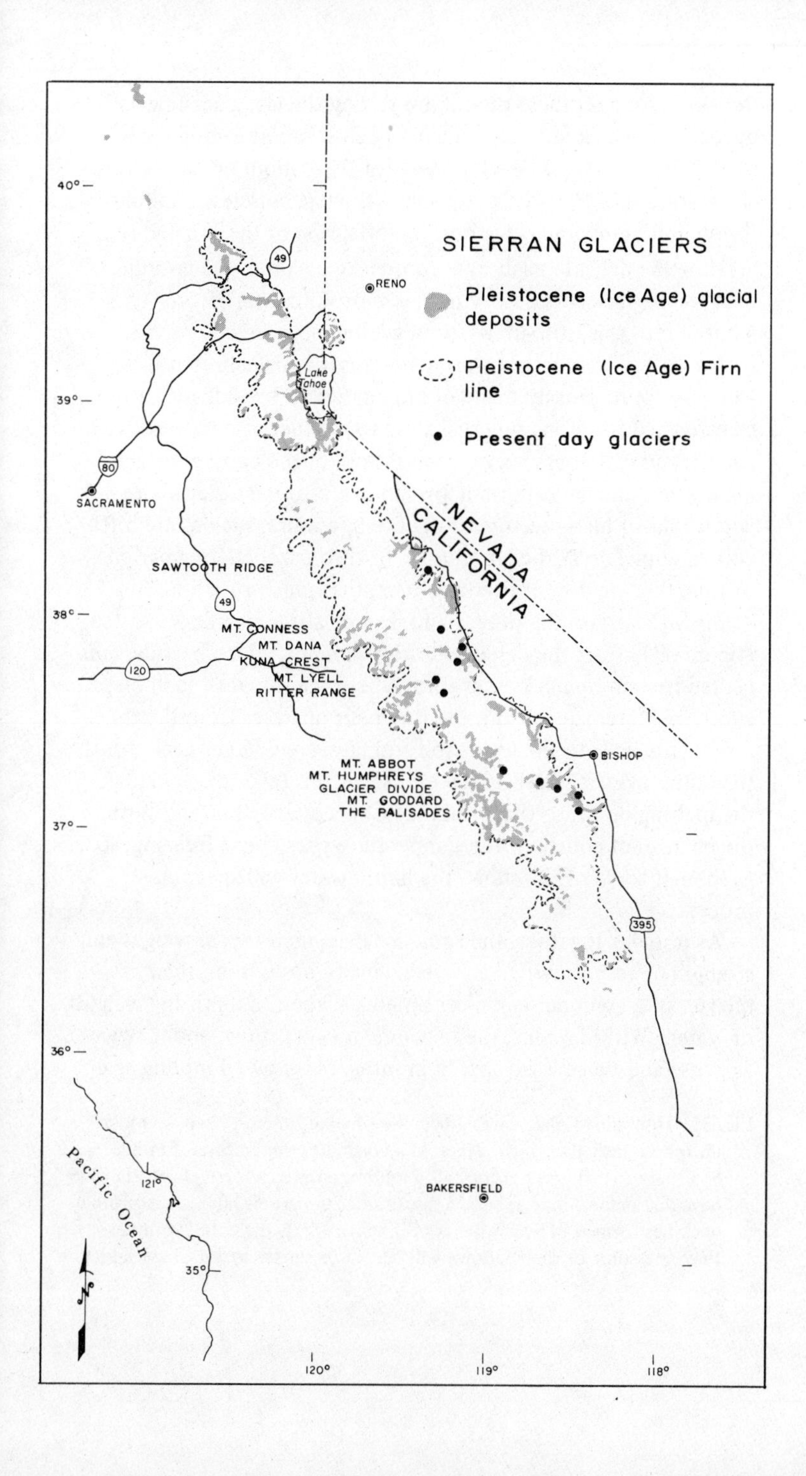
SIERRAN GLACIERS
Pleistocene (Ice Age) glacial deposits
Pleistocene (Ice Age) Firn line
Present day glaciers
40°
39°
38°
37°
36°
35°
121°
120°
119°
118°
49
80
120
395
RENO
Lake Tahoe
SACRAMENTO
NEVADA
CALIFORNIA
SAWTOOTH RIDGE
MT. CONNESS
MT. DANA
KUNA CREST
MT. LYELL
RITTER RANGE
BISHOP
MT. ABBOT
MT. HUMPHREYS
GLACIER DIVIDE
MT. GODDARD
THE PALISADES
BAKERSFIELD
Pacific Ocean
N

the work done on these mountains during the last glacial period by glaciers, which are only streams of closely compacted snow-crystals. . . . In the development [of these mountains] Nature chose for a tool . . . the tender snowflowers noiselessly falling through unnumbered centuries, the offspring of the sun and sea."

How the delicate, elaborate snowflake becomes the grinding blue ice of glaciers is a story in metamorphism that affords us a glimpse into this little-understood geologic process.

Ice differs from most other metamorphic rocks in that it is, whenever pure, wholly made of one material. In addition, it is transformed from one physical state to another at temperatures and pressures that are very close to those of the earth's surface. Most other minerals are transformed one into the other under conditions of high pressure or high temperature, such as are difficult to duplicate in the laboratory, or through such long periods of time that we do not live to witness the transformation.

Snowflakes, on the other hand, can be altered while we watch (figure 39). First they change to tiny spheres, which become connected by minute necks that form where the spheres touch one another. Water vapor from the tiny balls migrates from the surface of the balls to the necks to form larger and larger necks, until the entire mass is sealed together. All of this takes place below the melting point, so that water is not necessarily present during the change; of course, if the temperature rises above freezing, so as to melt some of the snow, the liquid water will speed the process.

As it melds together, the snow-ice mass increases in weight and strength. Fresh snowflakes have a density of no more than a quarter of a gram per cubic centimeter – about a tenth the weight of water. Within weeks, they become powder snow, about twice as dense and twice as strong; in months, old snow, doubling again

Fig. 39. How snowflakes, having become tiny ice spheres, begin to metamorphose into glacial ice. Here, at a constant temperature of minus 5° C., are two spheres of ice half a millimeter in diameter. In *A*, they have just made contact; in *B*, 5 hours and 20 minutes later, a discernible neck has formed between the two. Eventually, if the cold continues, they and some of their fellows will become a single crystal of glacial ice.

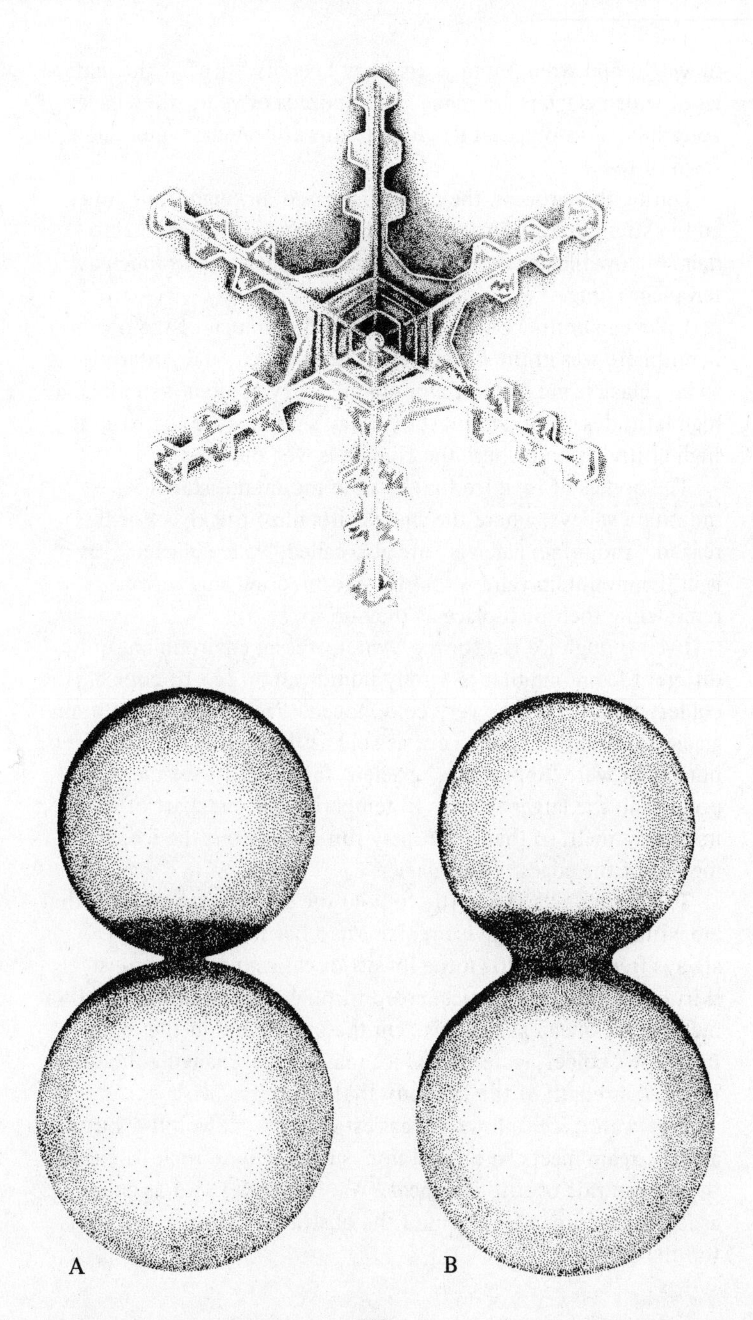
A
B

in weight and strength; in years, they become "firn" – the material of which glaciers are made. In hundreds of years, they have consolidated into glacial ice, nine-tenths as heavy as water and a form of rock.

During this process, they have increased in weight nine times and in strength 500 times. They have changed in shape from dainty snow filagrees to dense crystals of glacial ice as much as ten inches long.

Unless an entire continent is glaciated, as much of the northern hemisphere was in the Pleistocene glacial epoch, and Antarctica is today, glaciers are born in mountains. They abide not so much at high latitudes (although most of today's glaciers are there) as at high altitudes, and where the climate is wet and cold.

The bodies of rock ice that form in mountains start in mountain valleys, where the snow drifts most thickly. For this reason, "mountain glaciers" are also called "valley glaciers," as it is in the mountain valleys that they form, grow, and move, remodeling their birthplace as they do so.

Even though ice is a form of water, it is an environment quite different from one that is wholly liquid water. Ice, of course, is colder; polar ice can be very cold indeed. Probably the mountain glaciers of California were not as cold as the continental ice sheets, but rather were "temperate" glaciers, far closer to the melting point than the larger sheets. In temperate glaciers, part of the ice itself may melt, so that water may run throughout the ice, pouring out of the edges of the glacier.

The glacier acts as a kettle to hold the cold rivers and lakes that are within it. While the ice remains as a barrier, water is not always free to flow down the lowest or easiest path, but must skirt obstructions in the ice, creep through tunnels in the sheet, or melt the ice ahead as it flows. On the other hand, if the temperature grows colder, water in the ice may freeze, changing the texture and strength of the ice body that holds it.

Like water, ice follows the easiest course – downhill – but if an ice stream meets an obstruction, such as a hard rock barrier, the ice can ride over it as a sheet. Water, which is not as cohesive as ice, must form a lake behind the obstruction until it can overtop the dam.

As a stream of water is commonly deepest in the center and thinnest at the edges, so a glacier is thickest in the center. But, unlike water, the ice stream is usually bowed upward in the middle, rather than being nearly flat. For this reason, water on the surface of the glacier will run toward the sides as well as toward the general downhill slope, carrying with it pieces of rock and sand from the surface of the glacier.

Unlike rivers, most of which enter other bodies of water – last of all the sea – glaciers come to an abrupt end. The ice stream is more like a lake than a river in that it has geographic limits – a definite beginning and a definite end. Rivers rise from large, undefined, irregular areas where the rains fall, water collecting bit by bit into a recognizable stream; mountain glaciers head in spoon-shaped basins, and their limits are well marked.

Both ice and water streams move downhill in response to the pull of gravity. Water moves at a quick rate of many feet per second, while solid ice must temper its flow to inches per day. Nonetheless, glaciers can occasionally move much faster than that. One Alaskan glacier surged recently at a clocked rate of 300 feet per day, and a Russian glacier moved so fast in 1963 that a village at its foot had to be evacuated.

Not all of a glacier's movement is the result of melting and refreezing. Glaciers can move as solids, by plastic flow, each molecule gliding across its neighbor as the giant ice sheet shuffles along. This sort of movement takes place when the ice reaches a thickness of 100 to 150 feet.

The ability of the glacier to move as a solid and to use tools as it moves makes it a unique agent of erosion. When it melts, it leaves behind sculptures of pristine beauty. The tools it uses are simple: rocks and water. By freezing to the enclosing rock, the glacier can pluck – quarry – huge blocks, particularly in such highly jointed areas as the granitic reaches of the Sierra Nevada. Using these stones as teeth in immense files, the ice body can scrape the sides and bottom of its basin. It gathers any loose rock or blocks not held firmly in place as it moves along, using the accumulation as a ball mill not only to grind the rocks in the mill, but to scour out the valley itself. Smaller fragments of rock suspended in the ice or in the water within the ice act as sandpaper to polish the sides and bottom. The sand itself may be ground to powder (glacial flour) as it moves with the ice (figure 40).

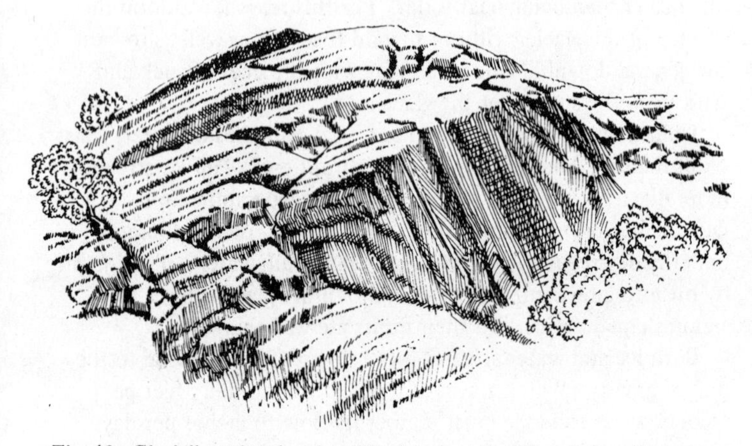

Fig. 40. Glacially striated and polished rock at Hamilton Lakes, Sequoia National Park. Each groove was worn by a stone or stones being pulled across it by the movement of glacial ice. Polished areas get their patina from the scouring action of minute particles of rock. Such polished areas glisten in the sunlight like old brass.

When a glacier has melted from its valley home, many evidences are left of its former passage. The valley has been transformed from the sharp-pointed, V-shape that it had when only a mountain stream ran through it, to a more gentle U-shape, reflecting the form of the now-gone ice. At its head, an amphitheatre will be left – a "cirque" – in which the only remnant of the glacier may be an ice-cold lake. In shape, this glacier head (oddly enough, the terminal, downhill end of a glacier is called its "snout") is somewhat like a tablespoon. It is generally deeper at the upper end than the lower, which is, of course, the reason why a lake is likely to remain there. The head wall above is very steep, but at the lip of the spoon there is often a ridge of bedrock, of heaps of broken rock, or a mixture of rock and ice. The Sierra's largest existing glacier, the Palisade near Bishop, has a heap of stones and ice forming a block at the end of the cirque.

There are about 70 tiny glaciers in the Sierra today, almost all, including the Palisade, cirque glaciers, living wholly within basins carved by the giant valley glaciers of the Ice Age.

Cirques are carved by active, dynamic glaciers. They are all fairly symmetrical, no matter what their orientation or what kind of rock they are carved in. They may be deeper and wider if the rock is soluble or is easily broken or already weathered, but their outlines are the same.

Since wind drift has a great deal to do with snow accumulation, cirque glaciers tend to form on the lee slopes of mountains. In most areas in the northern hemisphere, this means that northeast-facing slopes are favored. Palisade glacier hangs in such a niche, but a study of Sierran topography to locate cirques of the glaciers of the Ice Age would probably reveal many more cirques west of the main divide than east of it. They are there simply because more snow fell on the western side. The more symmetrical, larger ones are, nevertheless, those that lay longest in the shadow of the peaks rather than facing the heat of the summer sun.

The head wall, that part of the cirque that provides the most spectacular scenery, can stand thousands of feet high in resistant, jointed rocks like Sierran granite, while softer, less "competent" rocks crumble. Walcott cirque, near McMurdo Sound in the Antarctic, has a head wall nearly two miles high; Mt. Everest has one almost that high. The 1500-foot-high east face of Long's Peak, Rocky Mountain National Park, looks down on Chasm Lake, a tiny glacial tarn, resting in the cirque of an Ice Age sheet.

Most cirque glaciers have a crack around the ice – a "bergschrund" – parallel to the head wall. Palisade glacier, Mt. Dana, Mt. Conness, and Mt. Lyell glaciers all have well-developed ones. Just what the relationship of these cracks is to the erosive power of the glacier is not clear. In 1904, geologist W. D. Johnson descended the bergschrund on Mt. Lyell to its bottom, 150 feet below (figure 42). There he found the rock wall exposed, and discovered that "in the last twenty or thirty feet, rock replaced ice in the up-canyon wall. The schrund opened to the cliff foot. I cannot say that the floor there was of sound rock, or that it was level; but there was a floor to stand upon, and

Fig. 41. Three lakelets left as testimony of a former glacier. Lakes like these are called glacial step lakes, or, because they resemble beads on a rosary, are also known as paternoster lakes.

The three shown here may be seen toward the south from the top of Black Rock Pass, in the Mineral King quadrangle, Tulare County. They show very distinctly the effects of altitude on temperature: the highest lake, Columbine, is still frozen, and is connected to the next lake by a frozen waterfall. Below it, in the center of the drawing, is Cyclamen Lake, still frozen on its upper edges, but thawing toward the lower edge. It is connected to Spring Lake, the lowest, by a flowing waterfall. Ripples on Spring Lake show that it has completely melted. Step lakes are common in mountainous regions that have been sculptured by valley glaciers. Because the treads of the steps hold water and ice longer, they are more deeply eroded than the risers, gradually creating a lake basin.

not a steeply inclined talus. It was somewhat cumbered with blocks, both of ice and of rock; and I was at the disadvantage, for close observation, of having to clamber over these, with a candle, in a dripping rain, but there seemed to be definitely presented a line of glacier base, removed from five to ten feet from the foot of what was here a literally vertical cliff.

"The glacier side of the crevasse presented the more clearly defined wall. The rock face, though hard and undecayed, was much riven, its fracture planes outlining sharply angular masses in all stages of displacement and dislodgment. Several blocks were tipped forward and rested against the opposite wall of ice; others,

Fig. 42. Lyell Glacier, as it appeared in 1890. Its bergschrund, into which W.D. Johnson descended (see p. 131), is to the extreme right.

quite removed across the gap, were incorporated in the glacier mass at its base. Icicles of great size, and stalagmitic masses, were abundant; the fallen blocks in large part were ice-sheeted; and open seams in the cliff face held films of this clear ice. Melting was everywhere in progress, and the films or thin plates in the seams were easily removable."

Like rivers of water, glaciers coalesce to form larger ice streams. Unlike water, ice may fill its canyon to great depths, and yet come to an abrupt end. Sierran ice was as much as 4000 feet thick in some valleys at its greatest extent, and the entire ice cap was more than 100 miles long and 40 miles wide. Longest of the glaciers, Tuolumne, was fully 60 miles, but it terminated 30 miles from the foot of the range, at an elevation of about 2000 feet.

Where the ice was thick, small tributary glaciers in high mountain valleys met the main trunk at an altitude nearly that of the top of the ice sheet. While the ice was in the valley, an air view of the glacier would have looked like a wide sheet with branches. Now that the ice has melted, the difference in elevation between the trunk bottoms and tributary bottoms is striking. The base of the tributary glacier was, in some places, more than a

thousand feet higher than the base of the main trunk. Where a stream, now flowing in the glacially altered tributary, meets the main trunk, it drops breathtakingly to the main valley floor. Yosemite Falls, leaping in three sections over a 1430-foot-high precipice through a raceway 815 feet high, to a final 320-foot plunge to the floor of Yosemite Valley, for a total of 2565 feet, is such a "hanging" waterfall. So is Ribbon Falls, 1612 feet high – not all in free fall; and Bridal Veil, which plunges from a cliff 620 feet high. In little-visited Hetch-Hetchy Valley, Tueeulala Falls drops about 1000 feet, of which 600 are in free fall.

Most of the Ice Age mountain glaciers in the Sierra Nevada followed the courses of streams that had flowed in canyons before the great cold settled over them. The ice itself changed the high mountain portions of the stream valleys, and its melt-water altered the lower reaches.

By the time the Ice Age chill arrived, the distribution of land

Fig. 43. How a glacier transforms a mountain landscape. In the top view, rounded mountains with smooth slopes have many streamlets to drain them. Each little stream valley has a "V"-shaped profile, and meets the major stream at its own level. Deep soil, developed from rock through thousands or millions of years of weathering, covers the mountain slopes.

In the center view ice has formed in most of the valleys, and is slowly eroding the mountain slopes by quarrying and grinding. Long ridges of rock *(arêtes)* lead to a sharp-pointed peak ("matterhorn"), created by the erosive work of coalescing glaciers. To the right are small glaciers in horseshoe-shaped cirques. All of the glaciers in this drawing feed into the main glacier, which is flowing downward to the lower left. A line down its center marks a medial moraine. The tops of all glaciers meet, as the streams in the top view, at nearly the same level. What we cannot see, as it is covered by ice, is the difference in the elevation of the glacier bottoms – a difference that is due to the difference in ice thickness.

In the lower view, after the ice has retreated, a wholly different landscape has been created. The *arêtes* and the matterhorn are left as sharp ridges and peaks; the cirques, now bereft of glaciers, harbor only tiny lakes, or tarns. Step (paternoster) lakes now lead from the matter-horn down into the main valley. The streams that flow through them, and through the hanging valley to the left, finally meet the main gla-ciated valley as a waterfall.

and sea was close to that of today. The Sierra Nevada had become a rocky backbone; the Coast Ranges were separated from the Sierra by the Great Valley, no longer an inland sea. Since a great deal of water was locked up in ice during those times, the general level of the sea all over the world was lowered by about 300 feet.

On the western side of the Sierra, glaciers extended down the river valleys as much as 15 miles – quite a distance from the mountain source, but still a long way inland. None of the Sierran glaciers were close enough to the sea to enter the ocean directly, as polar glaciers do today. For this reason, and because the Coast Ranges were not ice covered, California does not have fjords like the Scandinavian countries, Canada, or Alaska.

Few of the glaciers on the eastern side of the Sierra extended beyond the mountain front. Those that did left remarkable evidence of their passing. Around Mono Lake, a shrunken remnant of one of a vast chain of Ice Age lakes, now-vanished glaciers have left ridges of rubble more than 800 feet high. In any setting other than between 9000-foot-high Mono Craters and the crest of the high Sierra, the ridges themselves would be regarded as mountains.

The ice made astonishing alterations in its valley – not so much in its lengthwise appearance, where the only noticeable change was to straighten curves slightly so as to make it easier for the ice to turn, but in the cross-sectional shape, where it sculptured the valley from a V-shape to a U. The U-shaped valley is a scenic wonder: the sharp, straight edges provide steep cliffs; the sudden drop permits spectacular waterfalls; the flattened bottom gives room for quiet lakes and marshes.

The world's most renowned example of such a valley is Yosemite. Its twin, Hetch-Hetchy, was once equally remarkable. Today, its glory is hidden. It is occupied by a lake, source of some of the water for San Francisco. Yosemite's valley has been called "nature's textbook on glacial erosion," for here one can see what ice, which at times filled the valley from rim to rim, can do (figure 44 and 45).

If you stand at Glacier Point and look toward Little Yosemite Valley, you can see how the valley has been cleared and reshaped. One stream of ice came from the east. It was joined by another from Tenaya Canyon just below Glacier Point, and together they created the Yosemite we see today.

Fig. 44. Bird's-eye view of Yosemite Valley as it was just before being invaded by glaciers of the Ice Age. The valley was deep, with a V-shaped inner gorge and side valleys. The entire region, up to timberline, was covered with coniferous forests, much as it is today.

In the center background is Half Dome, here a massive, cliffless mountain; in the foreground, left, El Capitan, not yet a sheer face. Cloud's Rest, left background, is only somewhat less eroded than today. Peaks on the skyline include Echo Peak, far left, and Mts. McClure, Lyell, and Florence on the right.

This reconstruction and the one following were drawn by the late François E. Matthes, student of Yosemite, who spent many years unravelling the geologic story of the valley.

Fig. 45. Bird's-eye view of Yosemite Valley as it probably was immediately after the Ice Age. The valley had been deepened and broadened; a lake 5½ miles long, dammed by a glacial moraine, occupied the valley floor. Large topographic features looked very much as they do today.

As the glacier moved slowly but relentlessly past Half Dome, it cleaned the dome face, carrying away pieces of granitic rock broken off along joint planes to leave a clean, steep precipice. At Glacier Point, the valley was completely filled with ice. So were the shallower side canyons feeding into it. When the ice melted, the side canyons were left stranded high above Yosemite Valley. The streams in them leap over the lip of the valley in spectacular waterfalls.

In the uplands near Tuolumne Meadows, the glaciers were not confined to narrow river canyons, but spread themselves out over broad reaches. Here you can see the work of "glacial *moulin*," or glacial mills, round holes formed by glacial waterfalls and eddies armed with scouring rocks. Here, too, you will see *roches moutonnées* ("rock sheep") – rounded rocks left on a deserted glacial floor (figures 46 and 47). The uphill side of these rocks rises smoothly and gradually, while the downhill, lee glacial, side is more irregular, steep, and plucked. As you can see, the direction of ice motion can be deduced from these rocks. Liberty Cap and Mount Broderick, at the mouth of Little Yosemite Valley, are two enormous *roches moutonnées.* They stood directly in the path of the glaciers and were overridden by them, yet they survived, each as a massive, unsubdued giant.

The other Yosemite domes are not *roches moutonnées,* as they were not overridden by ice; but the passing ice altered their shape, smoothing, polishing, and streamlining the upstream side, and quarrying and steepening the downstream side. Lembert Dome, Fairview Dome, and Pothole Dome were all well polished by ice; some of the polish has worn away, but the many remaining patches gleam in the afternoon sun.

Other small, telltale clues to the passing glaciers are scattered through the high mountains. There are "chatter" marks, crescent-shaped gouges made by ice pressure, similar to the marks on an arrowhead fashioned by steady pressure on hard rock. More revealing than many larger features, these tiny tokens point to the direction of ice movement. A particularly good set is at the foot of Mount Huxley, south of Yosemite in Evolution Basin.

Scratches, gouges, grooves, and other erosive indignities to the rock of the valley are not the only souvenirs of the vanished glaciers. The glacier adds to the landscape, as well as subtracting from it. Most of what it adds is the material it has torn or worn from its enclosing valley, broken into bits and carried to a new resting place. Sometimes large boulders ride down the valley in or on the glacier. When the ice melts, they find themselves a great distance from the outcrop that was their source, perhaps left high and iceless on a ridge. Such boulders are called "erratics" because they are strangers in the landscape. One may be a boulder of metamorphic rock standing like a monument in a field of granite; another may be a granite monolith marooned in metamorphic terrane. There are many such out-of-place boulders in the Sierra Nevada.

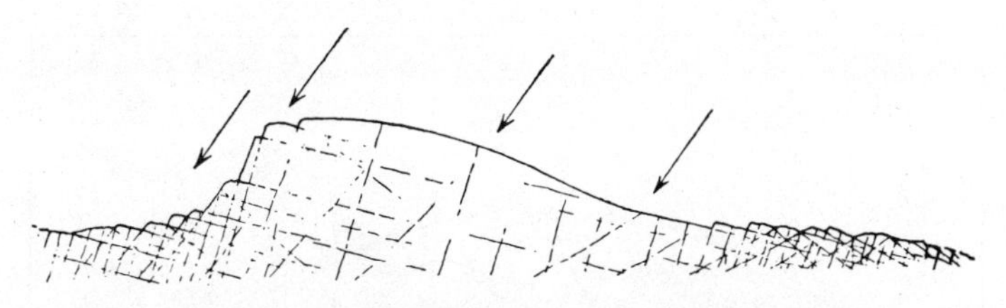

Fig. 47. Cross-sectional drawing showing how a glacier can fashion a *roche moutonneé* from "an obdurate mass of sparsely jointed granite. The glacier moved from right to left and exerted its force in the direction indicated approximately by the arrows – that is, at a high angle against the back and crown of the hump but at a slight angle away from the downstream face. It consequently subjected the back and crown to vigorous abrasion, leaving them smoothed and gently curved, and it subjected the downstream face to quarrying mainly, leaving it hackled and abrupt."

As the ice moves forward, it pushes small and large rocks ahead of it much as a snowplow pushes snow, to form a rocky ridge called a "moraine" (see figure 48). Some of the material the ice can actually plow up, but most rides along on top, beneath, or within the ice until it reaches the glacier's snout, where it joins other material being washed and pushed ahead of the glacier.

Rock and dust of various sizes find their way to the sides, too, to form other moraines. These, like the moraine at the snout, are a mixture of large rocks and small in no particular beds or arrangement. Moraine is the French word for "hill" or "rubble heap," indicating the unsorted arrangement of rock material in it.

Fig. 46. *Roches moutonneés* at Tuolumne Meadows, Yosemite National Park. On the skyline are Unicorn (left) and Cathedral Peaks (right).

The name *roche moutonneé* (rock sheep) was suggested in the eighteenth century by Horace Benedict de Saussure, a pioneer geologist. He, like F. E. Matthes who made the drawing shown in the above illustration, was a leader in the young sport of mountaineering. During one of his climbs, he recognized the glacial origin of this peculiar rock form. *"Roche moutonneé,"* he said, probably intending a pun: not only do these rocks, sprinkled through a mountain valley, resemble a flock of sheep from a distance, but *"moutonneé"* was the term used to describe a stylish wig, pomaded by mutton tallow, that was popular in the salon society of his day.

Fig. 48. How a glacier leaves heaps of rock debris as it melts back. Along the sides are sharp-crested ridges of rock – lateral moraines – curving around the end to form a series of terminal moraines. Each inner loop is younger than the outer one, as it marks the position of a retreating glacier. In the right foreground, a glacial stream has cut through the terminal moraines, carrying meltwater and a portion of the moraines away.

The word "moraine" is French for "rubble heap;" it is a heap of unsorted glacial material called "till," composed of rock debris of all sizes from angular blocks the size of automobiles to very fine rock flour.

In the United States, the Scottish word "till" is used to refer to the rocky debris of which the moraine is made.

The weather helps the glacier to break up the rock of the mountains. In the course of freezing and thawing 300 times, ten tons of dust may be derived from five acres of granite mountain-side by the breaking up of minerals in the rock. All of this, together with rock debris from avalanches, rock slides, falling boulders, and the like, drop onto the glacier to become part of the glacial moraines.

In many glaciers, the ice surface is thoroughly scored by crevasses. Crevasses may be less than an inch, or more than

50 feet wide. They may be a few feet to hundreds of feet long, and as much as 150 feet deep. They form where the glacier is under stress: where it turns a corner; where the central ice is travelling faster than the ice on the edge; where there are knobs in the floor of the valley. Rocks and dust fall into each of these cracks, to make their way through the ice as the glacier moves. When the glacier melts, the rock material that is still in the crevasses – has not yet made its way downward – is left as an unsorted ridge on the valley floor, forming winding lines among the scattered boulders dropped from the ice.

Glacial streams flow over the surface, along the bottom, and within the ice. They run through definite tunnels, rather than as a sheet along the bottom. Those in the tunnels within the ice form a pattern like the branches of a great tree, pulling water in toward a downhill trunk. When the confining ice collapses, the debris in the stream bed, like that in the crevasses, is dropped to the drying valley floor. Former glacial stream beds make sinuous ridges that mark the old stream courses. They are the reverse of stream beds on land: the ice that supported them has vanished, so that the sand and gravel once in the water is now left elevated. When a stream on land dries, the material in its bed is below the surface of the land, for the land itself, not ice, was the container.

Despite their vastness and unrelenting erosive power, glaciers are quite sensitive. Born of climatic change, glaciers reflect the slightest alteration in climate. When more snow falls in the winters than melts in the summers, a glacier may begin to form and to grow, pushing moraines ahead of it as it advances, travelling downhill both by growth and by movement of the ice particles. If the climate warms, or the snows decrease, the glacier may have insufficient nourishment to maintain its great size. Melting faster than it is being added to, the glacier is said to "retreat," leaving heaps of rocky debris behind it. It may abandon its end – terminal – moraine, as well as a series of "recessional" moraines that mark resting places in the general ebb.

The last Ice Age in North America and Europe was marked by several advances and retreats of the enormous ice sheets. These have been clocked by a number of methods, but just how their timetable matches that of the Sierran valley glaciers is not yet completely known. Study of the advances and retreats of the mountain ice cap glaciers of the Sierra Nevada has shown that there were many advances – "glaciations" – since the beginning of the Pleistocene Ice Age. Table 8 shows how they may relate to continental glaciations.

Since many of the moraines that are the clues to the various ice advances, retreats, and hesitations may be mixed together in one area, or else are so separated from one another that it is not easy to see exactly how the moraines match from one valley to another, reading the story of the Ice Age in the Sierra requires the delicate unravelling of many morainal strands. The strands have not yet been untangled.

It is rare that the age of a moraine can be measured directly. Once in a while, a splinter of wood caught in a younger moraine may be found, and the age of the wood calculated by carbon-14 methods. This gives the time when the wood was buried, and is a useful method for determining the age of moraines less than 40,000 years old.

More circumstantial evidence must be used to decipher most of the record. The weathering of rocks in the various moraines is one criterion used. So is the development of soil – a tool that is helpful in deciphering the story of continental glaciation, but soil is scarce on most Sierran glacial deposits. As an added complication, both the depth of weathering and the amount of soil are greater on the western side of the mountains than on the eastern, as there is much more moisture on the west to aggravate weathering and the production of soil.

Lava flows and faulting are also clues to the age of glaciation. For example, glacial debris lies on top of lava flows near Bishop. The flows have been dated by radiometric methods as 3.2 million years old. Since the glacial till lies above, it must therefore be less than 3.2 million years of age. Above the till, in turn, is an outcrop of the 700,000 year-old *nuée-ardente* deposit described on page 108. At this place, at least, we can say that this particular glacial till is more than 700,000 and less than 3.2 million years old. By tracing the till to other outcrops, it might be possible to estimate its true age more accurately.

Table 8

Major Sierran Glacial Advances

North American continental glaciation	Sierran glaciation	Approximate age in years[a]
Modern (Holocene) Time		
Neoglacial (began about 10,000 B.P.)	Matthes	700 B.P. to present
	Recess Peak	2000 to 2600 B.P.
	Climatic Optimum	
	Hilgard	9000 to 10,500 B.P.
Ice Age (Pleistocene) Time		
Wisconsin (began about 85,000 B.P. ± 15,000 years)	Tioga	Maximum ice about 20,000 B.P.
	Tenaya	Maximum ice about 45,000 B.P.
	Tahoe	Maximum ice about 60,000-75,000 B.P.
Illinoian (began about 115,000 B.P.)	Mono Basin	Maximum ice about 130,000 B.P.
Kansan (began about 400,000 B.P.)		
	Casa Diablo	Maximum ice about 400,000 B.P.
Nebraskan (older than 400,000 years[b])	Sherwin	Older than 700,000 years[c]
	McGee	Older than 2,600,000 years[c]
	Deadman Pass	2,700,000 to 3,100,000 B.P.[c]

a B.P. is before present; B.C., Before Christ; A.D., Anno Domini (years since birth of Christ)

b The beginning of glaciation in California has been dated at about 3 million years ago. This is much earlier than has been generally recognized as the beginning of continental glaciation in North America.

c Age of maximum ice, or age of entire glacial episode not yet known.

In some places, moraines have acted as barriers to lava flows, impounding or diverting them. Clearly, then, the till would have been in place before the lava. On the other hand, a lava flow or a volcano can be too high to be overridden by ice or covered by moraine; there is no doubt in such a case that the ice was younger than the volcano.

Volcanic ash is proving to be another valuable asset in reading the story of the last "few" years of the Sierra Nevada. In Tuolumne Meadows, Yosemite, for example, searchers found a lump of charcoal in an ash bed which proved, by carbon-14 methods, to be about 1550 years old. Since it is possible to measure the rate at which soil is accumulating on top of the ash, and since geologists have determined the age of the ash, it is possible to guess quite accurately when vegetation first began to grow in the meadow.

Another indication of by-gone climates – and, by inference, of glacial advances and retreats – is the fluctuation in the elevation of timberline. In the White Mountains, east of the Sierra proper, timberline in years past was determined by mapping, by tree ring counts, and by carbon-14 methods. The record correlates very nicely with glacial advances and retreats, thereby giving us still another check on the antiquity of past glaciers.

The story of the Ice Ages in the Sierra, as it now reads from such clues, shows that the Sierra harbored glaciers more than 3 million years ago – as long ago as any Ice Age glacier was formed anywhere. There were glacial advances about 400,000 years ago, and again about 130,000 years ago, both of which left their marks on the eastern Sierra. (See table 8.)

Best known of all the glacial deposits, because they stand out so sharply in the landscape, are the moraines left by the glaciers of 60,000 to 20,000 years ago. They are well developed on the eastern side of the Sierra, where they form low, voluptuous ridges most easily seen in the orange light of sunset. Little Walker Lake, above Mono Lake, is enclosed by moraines left by glaciers of four different glaciations (see table 10). Here, the moraines are fresh enough that one can almost see the cold, vanished giants that made them.

Many moraines are visible from U.S. Highway 395 between Reno and Lone Pine, and can be followed into the canyons from which they came. There are moraines to be seen on Mt. Rose, Nevada; around Lake Tahoe (Fallen Leaf Lake was formed by a morainal dam); near Bridgeport; along Lee Vining Creek, in Gibbs Canyon; in Mammoth, Sherwin, Laurel, Convict, Rock, McGee, Hilton, Pine and other canyons as far south as Olancha Peak.

In some places, if one is aware of the techniques for distinguishing older moraines, and knows where to look, he will be able to see those that are more than 100,000 years old (in Gibbs Canyon, for example); but for the most part, the moraines that are most visible are the long, trailing ridges of the Tioga-Tenaya-Tahoe glaciers. Moraines left in Tioga time mark the end of the great Ice Age. The mountain glaciers melted, and the continental ice sheets vanished.

But we are not finished with glaciation. In the last 10,000 years, the climate has cooled enough three times for small glaciers to form. Yosemite geologist François Matthes called the latest of these the "Little Ice Age," which now bears his name in honor of the work he did in deciphering the story of the Sierran landscape.

From a time of glaciation lasting from about 8500 B.C. to 7000 B.C. (9000-10,500 years Before Present – B.P.), the climate gradually warmed to the "climatic optimum" – an era of warmth that lasted through most of the days of Ancient Egypt to about 600 B.C. Then, as Greek and Roman civilization waxed and waned, the earth grew colder, up to about the time of Christ when warming commenced again. Had Hannibal been born a few centuries later, in an interglacial time, his elephants might easily have walked over the snowless Alpine passes. Rome, not Carthage, might have been forever destroyed.

Warmer times continued up to 1300 A.D. The northern seas were open. About 985, Eric the Red and a band of Norsemen founded colonies in Greenland that grew large and prosperous. From one of them, Leif the Lucky, Eric's son, set forth on an expedition that led him to the American continent, landing in about the year 1000. In the twelfth century, there were nearly 5000 colonists on Greenland, working 280 farms.

Then the drab cold days set in. Permafrost rose in the ground, freezing seeds and making even the digging of a grave for the dying impossible. Ice filled the harbor. Cattle died, and even wolves could find no food. Fog, wind, and torrential freezing rain swept the fated colony. In the gray mists that swirled over the whitening land, the last survivor perished.

The colony had come to a virtual end in 1410, when the last ship left. The history of Eric the Red and his colony passed, like Pompeii, into legend, a victim of changing geology. Two hundred years later, as the cold was reaching its maximum, the Pilgrims came to North America. Small wonder that they gave heartfelt thanks for having lived through that first bitter, glacial winter.

Little Ice Age alpine glaciers were at their height from about 1700 to 1750. Most of California's present glacierets were as large and strong then as they ever were. They are not descendants of the Ice Age glaciers of 20,000 years ago, but are new, infant glaciers that look small and out of place in the cirques carved by their powerful predecessors.

From 1750 to about 1850, the climate warmed again, melting part of the small glaciers. Then once more it cooled, and the tiny glaciers grew a little larger. In 1883, for example, there was a gla-cieret in Gibbs Canyon, where only a minute patch of ice remains today.

From the turn of the century until the 1960s, glaciers melted rapidly. About seventy small ones still hang in the cirques of the high country, or lie hidden beneath a protective rock covering. Northern glaciers have been expanding during this past decade; perhaps Sierran ones are also. (See table 9.)

Besides these tiny alpine ice glaciers, the Sierra has many "rock glaciers," mostly tracing their origin to this recent cold period. These are not true ice glaciers, because they are not composed principally of ice. They are made of broken rock, mixed with ice, but they move downhill as a wrinkled mass, much as a true glacier does.

Why did the earth grow cold? Although the Great Ice Age was not the only period of glacial cold in the earth's four-billion-year-long history, there were very few, if any, like it. More than fifty ideas as to why glaciers should cover so much of the earth have been suggested, none of which has been wholly satisfactory. These include, among many others, erupting volcanoes, throwing dust up to block out the sun's heat; shifts in the position of the earth's poles; and movement of continents. Perhaps the answer lies outside the earth, in astronomical causes. Perhaps the sun gave less heat, or the earth was differently tipped toward it.

Whatever the driving force that pushed the continental glaciers over mountain ridges and formed ice caps where there are none today, it was powerful. When it ceased, the glaciers waned, the ice sealed up in the ice caps melted, and the sea rose, filling the sea basins a little beyond their brims, separating lands that once were connected.

While Asia and America were connected, men crossed to the New World through the Siberian bridge. Some American horses, camels, tapirs, and zebras left the New World for Asia, while those that remained behind quickly became extinct. By the time men of the New and Old World were once again aware of each other's existence, all memory of American horses, camels, and zebras had vanished, although this had been their birthplace.

Will the earth cool again? Perhaps. We cannot now predict what nature will do. But we can guess what man can do. If the cloud cover, today about 31 percent of the earth's surface, were to be increased to 36 percent, it has been estimated that the average temperature would drop seven degrees F. – enough, according to some students of glaciation, to trigger a new ice advance. We are now fully capable of making such a change in the cloud cover, whether we intend to or not, and new glacial advance could mean worldwide catastrophe.

Table 9
Existing Glaciers in the Sierra Nevada

Glacier	Elevation of lower and upper margins in feet above sea level	Location
Sawtooth Ridge (many)	10,600-11,200	Yosemite National Park, Matterhorn Peak quadrangle
Mt. Conness (several)	11,000-12,000	Yosemite National Park, Tuolumne Meadows quadrangle
Mt. Dana	11,200-12,300	Yosemite National Park, Mono Craters quadrangle
Kuna Crest (several)	11,400-12,600	Yosemite National Park, Mono Craters quadrangle
Mt. Lyell (several)	11,500-12,800	Yosemite National Park, Merced Peak and Tuolumne Meadows quadrangle
Ritter Range (many)	10,500-12,200	Devils Postpile quadrangle
Mt. Abbot (several)	12,000-13,000	Mt. Abbot quadrangle
Mt. Humphreys (several)	11,400-12,800	Mt. Tom quadrangle
Glacier Divide (many)	11,400-13,000	Mt. Goddard and Blackcap Mountain quadrangles
Mt. Goddard (many)	11,400-12,800	Mt. Goddard quadrangle
The Palisades (many; some large)	11,200-13,000	Big Pine and Mt. Goddard quadrangles

Table 10

Features of Sierra Nevada Glaciers

Feature	How to recognize it	Probable origin	Where to see a good example
Bergschrund	Crack in ice parallel to head wall of glacier	Developed near head wall by meltwater trickling through head wall in relief of pressure	Dana (map 9, circle 12), Palisade (map 12, circle 13), Lyell (map 9, circle 39), Conness (map 9, circle 7), McClure (map 9, circle 38) (most Sierran glaciers)
Crevasse	Crack in ice not necessarily parallel to head wall	Developed where ice breaks in zones where one part is traveling faster than another or where there are obstructions beneath ice	Most Sierran glaciers
Rock glacier	Corrugated mass of angular rock near cirque, shaped like glacier	Mixture of ice and rock, moving like glacier, in areas not now cold enough or wet enough for glacier to form	Eastern side of Sierra Sherwin Canyon (map 11, circle 9) Mt. Tom (map 12, circle 6) Cirque between Two Teats and San Joaquin Mountain (map 9, circle 36) Mt. Gabb (map 11, circle 18)

Table 11

Glacial Features of the Sierra Nevada

Depositional

Feature	How to recognize it	Probable origin	Where to see a good example
Erratic	Markedly different type of rock lying in terrane not its source	Boulder carried in or on ice, left when ice melts	Chiquito Creek (map 10, circle 10) (granite on lava) Faith Valley (granite on lava)(map 6, circle 2) June Lake (map 9, circle 26) Rock Creek (map 12, circle 4) Balloon Dome (map 11, circle 14) Humphreys Basin (map 12, circle 8) Sentinel Dome (map 10, circle 23) Starr King Meadows (map 10, circle 7) Moraine Dome (one type of granite on another (map 10, circle 21) Glacier Point (metamorphic rock on granite)(map 10, circle 24) Cathedral Rocks (map 10, circle 26) Twin Lakes near Kaiser Peak (map 11, circle 19)(granite on calcareous rock) Bighorn Plateau (map 14, circle 1)

Feature	How to recognize it	Probable origin	Where to see a good example
Kame terrace	Mound of poorly sorted sand and gravel forming ridge along glacier edge	Deposited in channels of streams at edge of ice	Junction of Sonora Pass Highway (State Highway 120) and U.S. Highway 395 (map 6, circle 15)
Moraine Terminal Lateral Medial	Long hills of unsorted sand, gravel, clay, and boulders; ridge form is distinctive Some boulders may be faceted	Deposited at sides (lateral), in center (medial), or end (terminal) of glacier; medial most commonly forms where two glaciers join	Moraines enclose Walker Lake, at foot of Bloody Canyon (map 9, circle 19) Gibbs Canyon (map 9, circle 11) McGee Canyon (map 11, circle 13) Convict Lake (map 11, circle 11) Sawmill Creek (map 9, circle 20) Lee Vining Canyon (map 9, circle 10) Twin Lakes, Matterhorn Peak quadrangle (map 9, circle 2) Lateral moraine near State Highway 168, Huntington Lake 15-minute quadrangle (map 11, circle 21) Modern glaciers have several small terminal moraines: Kuna (6 or 7)(map 9, circle 23); Dana (3 or 4)(map 9, circle 12); Palisade (map 12, circle 13); Lyell (map 9, circle 39)

Feature	How to recognize it	Probable origin	Where to see a good example
Moraine-dammed lake	Has natural dam made of till	Water held in by ridge of till	Convict Lake (map 11, circle 11) Donner Lake (map 3, circle 1) Fallen Leaf Lake (map 3, circle 5) Gilmore Lake (map 3, circle 6) Grant Lake (map 9, circle 24) June Lake (map 9, circle 26) Walker Lake (map 9, circle 19) See also till
Outwash plain	More or less stratified deposit in valley beyond moraine	Deposited by glacial melt-water carrying fragments of eroded moraine	Sand Meadows (map 9, circle 15)
Perched boulder (see also glacial erratic) Also called glacial table	Pedestal of local rock capped by erratic May resemble mushroom	Local rock protected from weathering by more durable erratic	Starr King Meadows (map 10, circle 7) Upper Yosemite Fall (map 10, circle 14) Moraine Dome (map 10, circle 21) Chiquito Creek (map 10, circle 10), on trail to Chiquito Pass Parker Creek (map 9, circle 21)
Till	Jumbled mass of clay, sand, and boulders; some boulders may be 25 feet in diameter, some may	Deposited by glacial ice on bottom or in moraine	Moraines on east side of Sierra: near Bridgeport (map 9, circle 1), Reversed Creek (map 9, circle 25),

Feature	How to recognize it	Probable origin	Where to see a good example
	be faceted Distinguished from volcanic mud flow (*lahar*) by presence of large numbers of non-volcanic boulders		Convict (map 11, circle 11), McGee (map 11, circle 13), Rock (map 12, circle 4), Walker (map 9, circle 20), Pine (map 12, circle 14), Creeks In Bloody Canyon (map 9, circle 18) In Sawmill Canyon (map 9, circle 19) Around Lake Tahoe (map 3) Around Lake Mary (map 11, circle 7) Yosemite National Park (map 10) Most Sierran passes

Table 12

Glacial Features of the Sierra Nevada

Erosional

Feature	How to recognize it	Probable origin	Where to see a good example
Arête (grat, comb, knife-edged ridge are synonyms)	Steep, sharp rock ridge between adjacent cirques	Quarrying by glacier in cirque	North Palisade (map 12, circle 13) Mt. Humphreys (map 12, circle 7) Kaweah Crest (map 13, circle 6) Ritter Range (map 9, circle 40) Mt. McClure (map 9, circle 38) Mt. Lyell (map 9, circle 39) Septum between Mt.

Feature	How to recognize it	Probable origin	Where to see a good example
			Whitney (map 14, circle 4) and Mt. Russell (map 14, circle 3) South of Mt. LeConte (map 12, circle 12)
Avalanche chute	Slick, steep, U-shaped groove barren of vegetation	Snow avalanches, following same path	Bearpaw Meadows, High Sierra trail (map 13, circle 4) Sequoia National Park (map 13) Mt. Whitney (map 14, circle 4) Mt. Hitchcock (map 14, circle 5) Hamilton Lakes (map 13, circle 5)
Chain lakes (*see* Glacial stairway)			
Chatter mark	Crescentic gouges and fractures, deeper on downstream (down-ice) end	Impact pressure of ice	Mt. Huxley (map 12, circle 11) Evolution Basin, near Sapphire Lake (map 12, circle 10) Grant Lake (map 9, circle 24)
Cirque (*kar, cwm, botn, corrie, hoyo,* are synonyms in other languages)	Bowl-shaped depression in mountain side; generally backed by steep cliff	Scouring of glacial ice at head of glacier; plucking of head wall	Throughout Sierra, in high country
Cirque lake (tarn)	Lake in cirque (see cirque)	In cirque eroded by glacier Lake may be	Throughout high country Good examples in Humphreys Basin

Feature	How to recognize it	Probable origin	Where to see a good example
		final remnant of glacier, or may occupy older glacial cirque	(map 12, circle 8), Mt. Humphreys (map 12, circle 7), Kuna Crest (map 9, circle 16) Gold Lake, on Feather River (map 2, circle 1) Garnet Lake (map 9, circle 42) Thousand Island Lake (map 9, circle 41) Crystal Lake (map 13, circle 9) and Eagle Lake, near Mineral King (map 13, circle 10)
Col (pass)	Low saddle in glacial ridge opposite two cirques	Erosion by the heads of two glaciers, coalescing to destroy part of *arête*	Mono Pass, above Bloody Canyon (map 9, circle 17)
Comb (see *arête*)			
Cyclopean stairs (*see* Glacial stairway)			
Giant's kettle (*see* Glacial moulin work)			
Giant's staircase (*see* Glacial stairway)			

Feature	How to recognize it	Probable origin	Where to see a good example
Glacial *moulin* work (pot-hole, giant kettle)	Cylindrical holes in rock of glacier floor	Grinding of rock by boulder or pebble, in glacial eddies and vortices of water within ice	End of Tuolumne Meadows, Yosemite National Park (map 9, circle 14)
Glacial polish	Shiny surface on rock; if weathered, shine may be worn off in spots All minerals shine	Polishing by finely ground rock in glacial ice	Throughout high country Base of El Capitan (map 10, circle 12), Three Brothers (map 10, circle 13), Washington Column (map 10, circle 17), Union Point (map 10, circle 25) and walls near Mirror Lake (map 10, circle 18), Yosemite National Park Hamilton Lake (map 13, circle 5) Mokelumne Wilderness (map 5B, circle 3) Kaiser Ridge (map 11, circle 20) Evolution Basin (map 12, circle 9) Kern River Canyon (map 13, circle 11) Bloody Canyon (map 9, circle 18) Recess Peak (map 11, circle 17) Lembert Dome, Tuolumne Meadows, Tioga Pass road (map 9, circle 13)

Feature	How to recognize it	Probable origin	Where to see a good example
			Big Arroyo (map 13, circle 8) Upper Merced Canyon, above Nevada Fall, (map 10, circle 22), back of Liberty Cap (map 10, circle 20), Mt. Broderick (map 10, circle 19), Tenaya Canyon (map 10, circle 6), Pywiack Dome (map 10, circle 3), Yosemite National Park
Glacial stairway (glacial staircase, cyclopean stairs, giant's staircase, glacial step; paternoster lakes are lakes in glacial stairway)	Series of flattish valley areas, commonly with lakes (paternoster lakes, chain lakes, glacial step lakes) connected to one another by steep areas commonly with waterfalls U-shaped valley and glacial marks indicate glaciation	Scouring of glacial bed	Yosemite Valley (map 10) Black Rock Pass (map 13, circle 7) Faith (map 6, circle 2), Hope (map 6, circle 1), and Charity (map 6, circle 3) valleys Sixty-lake basin (map 13, circle 2)
Glacial step lakes (*see* Glacial stairway)			
Grat (see *Arête*)			
Grooves (*see* Scratches)			
Hanging valley	Tributary valley much higher than main valley, usually marked	Thinner tributary glacier met main glacier at	Yosemite Valley, many falls (map 10) Hetch-Hetchy Valley, Tueeulala

Feature	How to recognize it	Probable origin	Where to see a good example
	by steep cliff, perhaps waterfall (hanging waterfall)	level where tops were even, bottoms not; when ice melted, channel of tributary glacier left much higher	Falls (map 10, circle 1) Evolution Basin (map 12, circle 9)
Hanging waterfalls (*see* hanging valley)			
Knife-edged ridge (see *Arête*)			
Matterhorn (horn)	Pyramidal, steep mountain peak in glaciated area	Erosion at the heads of three or more converging glaciers	Matterhorn Peak (map 9, circle 3) Mt. Huxley (map 12, circle 11)
Pass (*see* Col)			
Paternoster lakes (*see* Glacial stairway)			
Polish (*see* Glacial polish)			
Pothole (*see* Glacial *moulin* work)			
Roche moutonée	Rocky outcrop in glacial landscape, rounded on one side, irregular on other	Erosion by overriding ice, smoothing upstream side by abra-	Tuolumne Meadows (map 9, circle 14) Liberty Cap (map 10, circle 20) Mt. Broderick (map

Feature	How to recognize it	Probable origin	Where to see a good example
		sion, plucking (quarrying) downstream side	10, circle 19) Desolation Valley (map 3, circle 8) Glen Alpine Valley (map 3, circle 7) Blaney Meadows (map 11, circle 22)
Scratches, grooves	Long lines or indentations in rocks	Abrasion of glacial walls and floor by rocks embedded in ice	Evolution Basin (map 12, circle 9) Yosemite Valley (map 10) Bloody Canyon (map 9, circle 18) Hamilton Lake (map 13, circle 5) Cathedral Pass (map 10, circle 5) Royal Arches (map 10, circle 15) Kern Canyon (map 13, circle 11) Grant Lake (map 9, circle 24)
Tarn (*see* Cirque lake)			
U-shaped valley	Has horseshoe shape in cross section	Eroded by glacier, usually modifying stream valley	Yosemite Valley (map 10) Hetch-Hetchy Valley (map 10, circle 2) Kern River Canyon (map 13, circle 11) Evolution Basin (map 12, circle 9) Pine Creek Canyon (map 12, circle 14)

Chapter 10

THE MOUNTAINS TREMBLE

Why does the Sierra Nevada exist? Why are mountains where they are? When and how were they created? These are proper questions to ask of geology. In the answers lies the key to the riddle of the earth.

Geology is young, historically speaking, and men are young, geologically speaking. We are far from knowing the ultimate answers. What we can know are immediate answers, and as rapidly as we have new facts and new answers, we can adjust our ideas to fit them, pushing the boundary of the unknown back a bit further.

Some of the theories we have today may seem as ridiculous tomorrow as the idea that earthquakes are caused by a great turtle shaking the earth on its back. Such an idea of the earth was once held by millions of people. We call it primitive; yet our current theories may be just as far from the truth.

We know more about when the Sierra Nevada came to be than why. From about 210 to 80 million years ago, the granite core of the Sierra was formed and brought to the surface through uplift and erosion. Then from about 80 to 55 million years ago, great quantities of overlying rock, as well as part of the granite itself, were washed into the adjacent sea to form thick sedimentary beds in what is now the Central Valley and farther west in areas now removed from the Sierra. Many clues point to this conclusion, but two major ones are the key. First, there are no sedimentary beds 80 to 55 million years old in the mountains to show that deposition was going on throughout the Sierra; in fact, in most places, there are deeply weathered soils, indicating that erosion, not deposition, was the order of the day. Second, the rocks that were deposited in nearby areas have a chemical and mineralogical composition – particularly the feldspar minerals – that is what one would expect from rocks of the composition of the granite of the Sierra Nevada.

The mountains were worn to low hills by 55 million years ago (end of early Eocene Epoch). From that time to about 30 million years ago (Oligocene Epoch), events moved slowly. The low mountains contributed some gold to the Tertiary rivers, but little sediment to the sea.

When the volcanoes began erupting in earnest in the middle Oligocene, the pace of change increased. Rhyolite flows poured over the crest and rhyolitic ash blew into the river channels. Lava and mud flows followed, heaping up as much as half a mile in the northern Sierra. Some time during the volcanic episode (which we are still in), possibly 20 million years ago but certainly by 10 million years ago, the northern Sierra, and probably the southern as well, was uplifted and tilted to the west. Evidence for this are the abandoned stream channels lying high on the ridges, coupled with the deep canyons in which the present rivers run. We know that most of the canyons on the west were cut before glaciation began 3 million years ago, because their channels are partly glaciated and contain some debris abandoned by the glacier but not yet reworked by the streams. Perhaps all of the uplift was completed by the end of glacial times; perhaps it is still going on.

How were the mountains uplifted? It is probable that mountains rise slowly, in response to gravity and other earth forces. Certainly, in our perspective, they seem to have risen slowly in the geologic past. Whether this is actually true, or whether we simply do not have enough information to know whether they rose imperceptibly (in the human sense) or as a series of perceptible jolts, we cannot say.

In the past hundred years engineers have worked steadily at remeasuring mountains to see if they have risen. Some seem to be taller than they were, in spite of erosion. However, our measurements are so few, our instruments so crude, and our history so limited that we cannot be sure that direct measurement is accurate. But, even if our techniques are adequate, and our time span long enough, one must remember that measurements are made relative to sea level. The sea can and does change level, in response to earth forces and in response to the quantity of water available. A change in climate, melting or increasing the polar ice caps, can drastically alter sea level. So, perhaps mountains generally rise slowly; likely they do, but we must now give the verdict: not proven.

We know some mountains rise quickly, in human terms. Parícutin rose in a few months; Surtsey was born in weeks; acres of land have been added to Hawaii while tourists watched. On February 9, 1971, the San Gabriel Mountains outside Los Angeles rose six feet in a few moments in a moderate earthquake.

The Sierra Nevada has young volcanoes that rose quickly, and numerous faults that have lifted mountains and shaken the earth. There are old faults in the Sierra, such as the Bear Mountain fault zone, and the Melones fault zone in the Mother Lode, that have raised the mountains in days gone by, and have served as avenues for gold-bearing solutions. As far as we know, their "activity" – that is, when earthquakes were taking place along them – is long past: they seem now to be fossil earthquake faults, though this may not be true, for nature has many surprises.

Earthquakes are rarer in the Sierra Nevada than in the Coast Ranges, but the numerous active faults in the Sierra testify to the continuing rise of the mountains. Figure 49 shows some of the major fossil faults, together with faults that are known to have been active recently.

Faults are flaws in the continuous skin of rocks in the crust of the earth, along which the earth has moved. On maps they are shown as lines; if you see a fault in nature, it is likely to look like a cliff (if it has not been eroded away), or a low place, perhaps studded by a series of ponds. Rivers that go in unexpected directions may mark faults; so may topography that looks out of place in the landscape, or rock strata that do not connect.

The relationship between faults and earthquakes has only recently been thoroughly established. After the San Francisco earthquake of 1906, Henry Fielding Reid suggested that the earthquake is the result of "elastic rebound." As the earth builds up stress – for whatever reason – in zones of weakness in the earth's crust, it climaxes at the point where the rock will break suddenly. Since rock is somewhat elastic, it reacts like a rubber band that has been stretched: it breaks and "twangs." The vibration – the "twang" – is the earthquake.

Faults along the east side of the Sierra are the most active in the range. About 10 million years ago, faulting commenced to lift the mountains; by 2 million years ago, the work was essentially completed in the northern part, to make the mountains as they are today. Faulting in the south half of the range started later, and is not yet finished.

There are many active faults in the Sierra that are potential "earthquake" faults (see figure 49). In fact, the most severe earthquake California has had in historic time took place along the east face of the Sierra Nevada, destroying every brick, stone, and adobe building in the Owens Valley town of Lone Pine, and killing a tenth of the population.

On a frosty March night in 1872, the entire southern Sierra felt a shock that was as strong, if not stronger, than the San Francisco 'quake of 1906. Twenty-seven people died, most of them instantly. Sixteen were buried in one common grave, now marked by a monument. Most of them had no known relatives within reach, as they came from Ireland, Chile, France, Mexico, and eastern United States. It was – or should have been – a lesson in building construction in earthquake country. Fifty-two of the 59 buildings in the town of Lone Pine were destroyed, including *every* building not made of wood. Every dead or injured person was within one of the destroyed buildings. The only frame building to be severely damaged was a "cheap, unsubstantial shed."

John Muir, the Sierra's most eloquent lover, was living in Yosemite Valley on the other side of the Sierran crest at the time of the earthquake. Here is his description of it, perhaps the only eye-witness description of the effect of earthquakes on the mountains themselves:

> The shocks were so violent and varied, and succeeded one another so closely, that I had to balance myself carefully in walking as if on the deck of a ship among waves, and it seemed impossible that the high cliffs of the Valley could escape being shattered. In particular, I feared that the sheer-fronted Sentinel Rock, towering above my cabin, would be shaken down, and I took shelter back of a large yellow pine, hoping that it might protect me from at least the smaller outbounding boulders. For a minute or two the shocks became more and more violent –

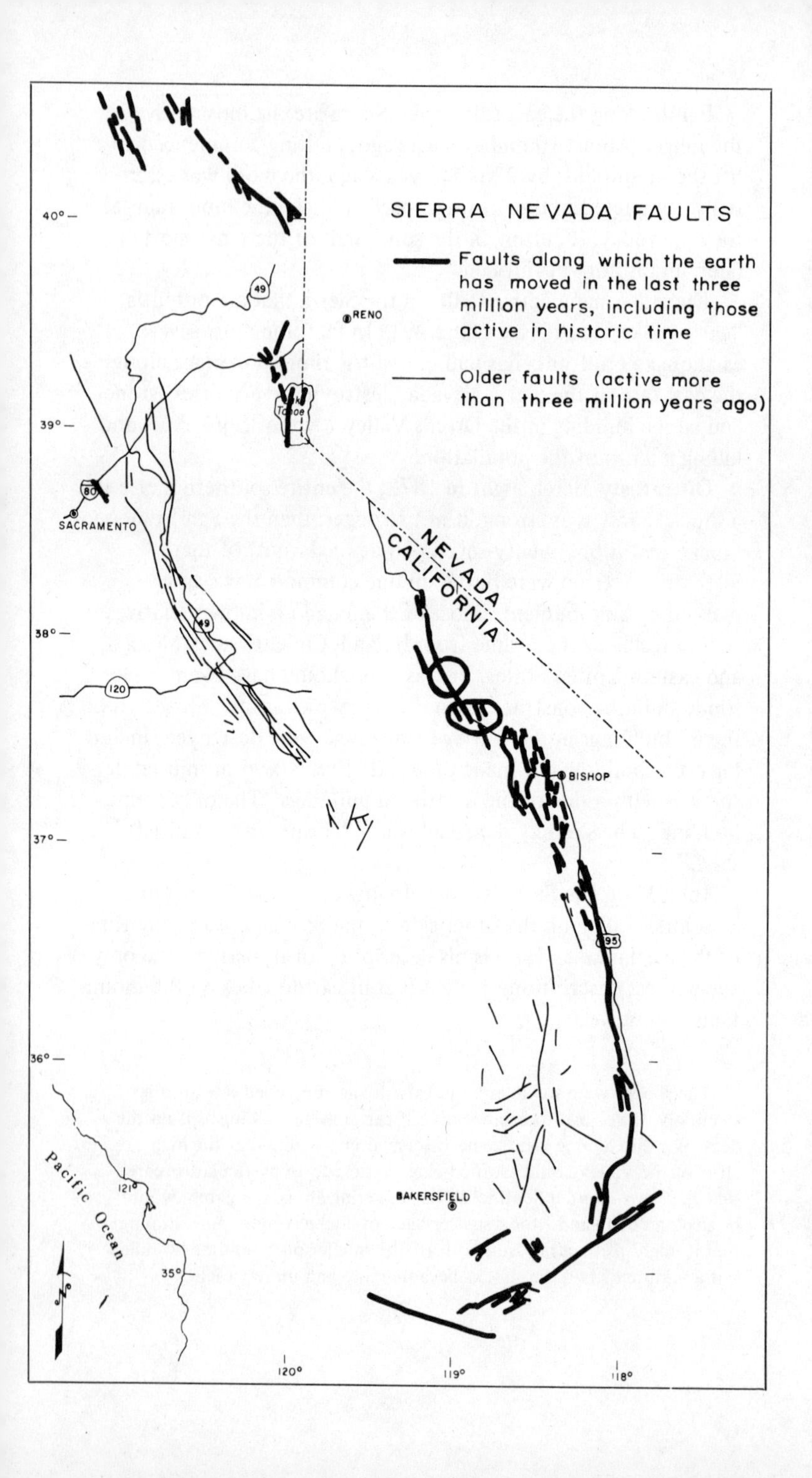

SIERRA NEVADA FAULTS
Faults along which the earth has moved in the last three million years, including those active in historic time
Older faults (active more than three million years ago)
40°
39°
38°
37°
36°
35°
120°
119°
118°
121°
RENO
SACRAMENTO
BISHOP
BAKERSFIELD
Tahoe
NEVADA
CALIFORNIA
Pacific Ocean
49
80
120
95

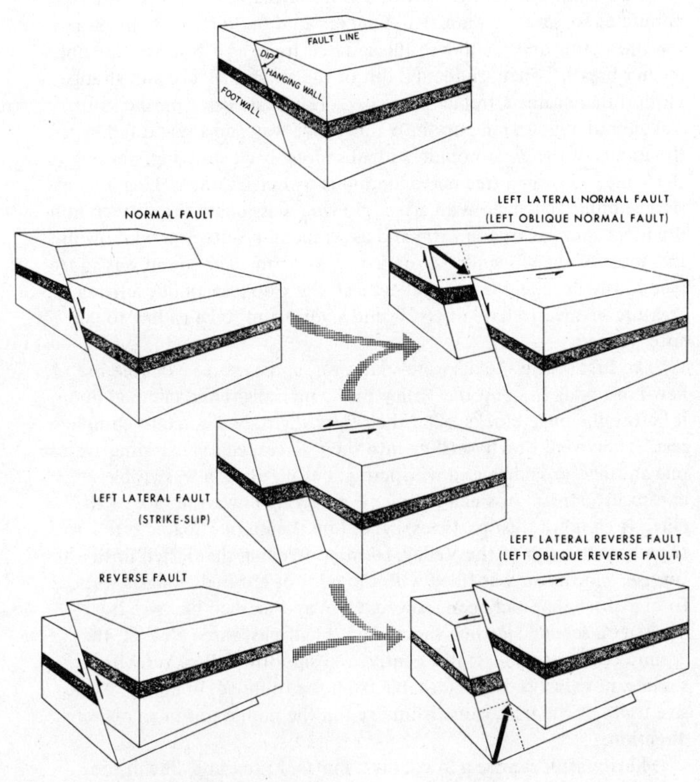

Fig. 50. Types of fault movement.

Fig. 49. Some of the faults in the Sierra Nevada. Those with light lines are "fossil" faults – those that raised the mountains in years past, but are now inactive, so far as we know. Heavy lines mark those faults that have moved since the beginning of the Ice Age, and may possibly move again. The ill-famed San Andreas fault passes by the Sierra Nevada at the most southern point of the range. At that point, the Garlock fault, southern boundary of the Sierra Nevada, joins the San Andreas.

flashing horizontal thrusts mixed with a few twists and battering, explosive, upheaving jolts, – as if Nature were wrecking her Yosemite temple, and getting ready to build a still better one.

It was a calm moonlight night, and no sound was heard for the first minute or so, save low, muffled, underground bubbling rumblings, and the whispering and rustling of the agitated trees, as if Nature were holding her breath. Then, suddenly, out of the strange silence and strange motion there came a tremendous roar. The Eagle Rock on the south wall, about a half a mile up the Valley, gave way and I saw it falling in thousands of the great boulders I had so long been studying, pouring to the Valley floor in a free curve luminous from friction, making a terribly sublime spectacle – an arc of glowing, passionate fire, fifteen hundred feet span, as true in form and as serene in beauty as a rainbow in the midst of the stupendous, roaring rock-storm. The sound was so tremendously deep and broad and earnest, the whole earth like a living creature seemed to have at last found a voice and to be calling to her sister planets. . . .

The first severe shocks were soon over, and eager to examine the new-born talus I ran up the Valley in the moonlight and climbed upon it before the huge blocks, after their fiery flight, had come to complete rest. They were slowly settling into their places, chafing, grating against one another, groaning, and whispering; but no motion was visible except in a stream of small fragments pattering down the face of the cliff. A cloud of dust particles, lighted by the moon, floated out across the whole breadth of the Valley, forming a ceiling that lasted until after sunrise, and the air was filled with the odor of crushed Douglas spruces from a grove that had been mowed down and mashed like weeds.

After a second startling shock, about half-past three o'clock, the ground continued to tremble gently, and smooth, hollow rumbling sounds, not always distinguishable from the rounded, bumping, explosive tones of the falls, came from deep in the mountains in a northern direction.

Shortly after sunrise a low, blunt, muffled rumbling, like distant thunder, was followed by another series of shocks, which, though not nearly so severe as the first, made the cliffs and domes tremble like jelly, and the big pines and oaks thrill and swish and wave their branches with startling effect.

During the third severe shock [that same morning] the trees were so violently shaken that the birds flew out with frightened cries. In particular, I noticed two robins flying in terror from a leafless oak, the branches of which swished and quivered as if struck by a heavy battering-ram. Exceedingly interesting were the flashing and quivering of the elastic needles of the pines in the sunlight and the waving up and down of the branches while the trunks stood rigid.

> It was long before the Valley found perfect rest. The rocks trembled more or less every day for over two months, and I kept a bucket of water on my table to learn what I could of the movements. The blunt thunder in the depths of the mountains was usually followed by sudden jarring, horizontal thrusts from the northward, often succeeded by twisting, upjolting movements. More than a month after the first great shock, when I was standing on a fallen tree up the Valley near Lamon's winter cabin, I heard a distinct bubbling thunder from the direction of Tenaya Canyon. The air was perfectly still, not the faintest breath of wind perceptible, and a fine, mellow, sunny hush pervaded everything, in the midst of which came that subterranean thunder. Then, while we gazed and listened, came the corresponding shocks, distinct as if some mighty hand had shaken the ground. After the sharp horizontal jars died away, they were followed by a gentle rocking and undulating of the ground so distinct that Carlo [the dog] looked at the log on which he was standing to see who was shaking it. It was the season of flooded meadows and the pools about me, calm as sheets of glass, were suddenly thrown into low ruffling waves.

The eastern mountain front rose upward at least 13 feet in a few moments that bright night, and moved horizontally perhaps 20 feet. Calculations – admittedly based on far too short an historic record – indicate that an earthquake of this magnitude might be expected along the Sierran front every few hundred years. Discounting smaller ones, and assuming that this is a "500-year" earthquake (on present evidence, this seems conservative), the rate of Sierran uplift from faulting, accompanied by large earthquakes, would be 26 feet per thousand years. At this rate, in a million years, the total uplift would be about 26,000 feet. True, 26,000 feet may be a great deal more than the Sierra Nevada actually rose in a million years, but it shows how much earthquakes can accomplish.

It is quite likely that much of the earthquake's energy was not translated into vertical uplift of the mountains. Owens Valley, perhaps, was dropped down; although this gives the same effect – it makes the mountains relatively taller – it does not lift them higher above sea level, which is what their altitude is based upon.

Field measurements of uplift in the last 3 million years – since the beginning of the Ice Age – based upon elevation of glacial moraines, height of cliffs ("scarps") produced by faulting, displacement of volcanoes and volcanic deposits, elevation of volcanic flows, and geophysical measurements, indicate that in the southern Sierra Nevada the mountains have risen or the valleys have dropped by virtue of faults and earthquakes perhaps as much as 19,000 feet. Much of the change has taken place since the hot ash fall of 700,000 years ago.

In the northern part of the Sierra, faulting has not been as prominent a means of raising the mountains. Instead, the northern part appears to be lifted up and tilted west, with far less fault breakage. The tilting, as nearly as can be determined on present evidence, was completed before the faulting in the southern part started in earnest.

HORRORS !!

Appalling Times!

EARTHQUAKES

Awful Loss of Life

25 Persons Killed !

EARTH OPENS !!

LONE PINE !

Fig. 51. Headlines as they appeared in the *Inyo Independent* following the earthquake of 1872. Altogether, twenty-seven people were killed in the Owens Valley area.

The northern mountain mass commenced to rise and to be tilted at the rate of about 90 feet per mile during the period from 25 to 12 million years ago, accelerating to 140 feet per mile in the span from 12 to 3 million years ago. At the end of the Pliocene Epoch, 3 million years ago, the southern portion began to be bent and broken, a process that is still going on, perhaps more slowly, today.

Looking backward into time, it is easier for us to draw arbitrary lines in history. "This is a period of uplift," we might say, "extending for three million years." Yet we do not know how many individual episodes of uplift there were within that time span, nor can we tell how many lifetimes might have been lived without any perceptible change in the mountains.

Today, the west coast is so shaky, with "earthquakes in divers places," that we do know we are living in a period of strenuous mountain making. The many shakes, terrifying though they may be to us who must live or die with them, are nevertheless creating the scenery we value so highly. Had we lived in a century with few or no earthquakes, we might have thought all activity had ceased; yet from a longer perspective, it is obvious that we live in the midst of violent times.

Geologists in the first half of the century saw pulses in the uplift of the Sierra Nevada that were separated by periods of erosion, in which the mountains were worn down to a more or less flat plain ("peneplain") close to sea level. Various geologists recognized three or four peneplains, and hence four or five stages of uplift, dating from the Miocene, 20 million years ago, to the Ice Age.

The facts they used were principally physiographic, based on the idea that large flat areas at approximately the same elevation were once continuous levels. Since continuous levels were thought to be produced only in the so-called "old age" of the landscape, when the mountains had been reduced to plains and there were no rushing torrents, steep canyons, sharp cliffs, or waterfalls, the flat areas represented times when the Sierra had been subdued. That there were three or four levels (ranging in elevation from the crest of Mt. Whitney to about Yosemite Valley) meant that there were several periods of uplift between.

THE MERCALLI INTENSITY SCALE

(As modified by Charles F. Richter in 1956 and rearranged)

If most of these effects are observed	*then the intensity is:*
Earthquake shaking not felt. But people may observe marginal effects of large distance earthquakes without identifying these effects as earthquake-caused. Among them: trees, structures, liquids, bodies of water sway slowly, or doors swing slowly.	*I*
Effect on people: Shaking felt by those at rest, especially if they are indoors, and by those on upper floors.	*II*
Effect on people: Felt by most people indoors. Some can estimate duration of shaking. But many may not recognize shaking of building as caused by an earthquake; the shaking is like that caused by the passing of light trucks.	*III*
Other effects: Hanging objects swing. *Structural effects:* Windows or doors rattle. Wooden walls and frames creak.	*IV*
Effect on people: Felt by everyone indoors. Many estimate duration of shaking. But they still may not recognize it as caused by an earthquake. The shaking is like that caused by the passing of heavy trucks, though sometimes, instead, people may feel the sensation of a jolt, as if a heavy ball had struck the walls.	*V*

If most of these effects are observed	*then the intensity is:*
Effect on people: Difficult to stand. Shaking noticed by auto drivers. *Other effects:* Waves on ponds; water turbid with mud. Small slides and caving in along sand or gravel banks. Large bells ring. Furniture broken. Hanging objects quiver. *Structural effects:* Masonry D*.heavily damaged; Masonry C* damaged, partially collapses in some cases; some damage to Masonry B*, none to Masonry A*. Stucco and some masonry walls fall. Chimneys, factory · stacks, monuments, towers, elevated tanks twist or fall. Frame houses moved on foundations if not bolted down; loose panel walls thrown out. Decayed piling broken off.	*VIII*
Effect on people: General fright. People thrown to ground. *Other effects:* Changes in flow or temperature of springs and wells. Cracks in wet ground and on steep slopes. Steering of autos affected. Branches broken from trees. *Structural effects:* Masonry D* destroyed; Masonry C* heavily damaged, sometimes with complete collapse; Masonry B* is seriously damaged. General damage to foundations. Frame structures, if not bolted, shifted off foundations. Frames racked. Reservoirs seriously damaged. Underground pipes broken.	*IX*

Other effects: Hanging objects swing. Standing autos rock. Crockery clashes, dishes rattle or glasses clink.

Structural effects: Doors close, open or swing. Windows rattle.

VI

Effect on people: Felt by everyone indoors and by most people outdoors. Many now estimate not only the duration of shaking but also its direction and have no doubt as to its cause. Sleepers wakened.

Other effects: Hanging objects swing. Shutters or pictures move. Pendulum clocks stop, start or change rate. Standing autos rock. Crockery clashes, dishes rattle or glasses clink. Liquids disturbed, some spilled. Small unstable objects displaced or upset.

Structural effects: Weak plaster and Masonry D* crack. Windows break. Doors close, open or swing.

VII

Effect on people: Felt by everyone. Many are frightened and run outdoors. People walk unsteadily.

Other effects: Small church or school bells ring. Pictures thrown off walls, knicknacks and books off shelves. Dishes or glasses broken. Furniture moved or overturned. Trees, bushes shaken visibly, or heard to rustle.

Structural effects: Masonry D* damaged; some cracks in Masonry C*. Weak chimneys break at roof line. Plaster, loose bricks, stones, tiles, cornices, unbraced parapets and architectural ornaments fall. Concrete irrigation ditches damaged.

X

Effect on people: General Panic.

Other effects: Conspicuous cracks in ground. In areas of soft ground, sand is ejected through holes and piles up into a small crater, and, in muddy areas, water fountains are formed.

Structural effects: Most masonry and frame structures destroyed along with their foundations. Some well-built wooden structures and bridges destroyed. Serious damage to dams, dikes and embankments. Railroads bent slightly.

XI

Effect on people: General panic.

Other effects: Large landslides. Water thrown on banks of canals, rivers, lakes, etc. Sand and mud shifted horizontally on beaches and flat land.

Structural effects: General destruction of buildings. Underground pipelines completely out of service. Railroads bent greatly.

XII

Effect on people: General panic.

Other effects: Same as for Intensity X.

Structural effects: Damage nearly total, the ultimate catastrophe.

Other effects: Large rock masses displaced. Lines of sight and level distorted. Objects thrown into air.

* Masonry A: Good workmanship and mortar, reinforced designed to resist lateral forces.
Masonry B: Good workmanship and mortar, reinforced.
Masonry C: Good workmanship and mortar, unreinforced.
Masonry D: Poor workmanship and mortar and weak materials, like adobe.

Fig. 52. The modified Mercalli Intensity Scale.

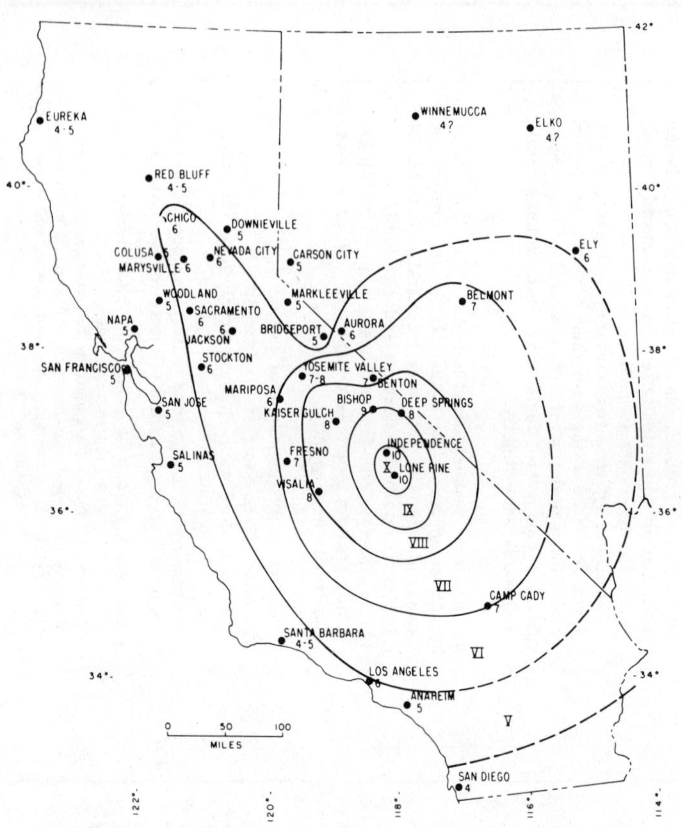

Fig. 53. "Isoseismal" map of the Owens Valley earthquake of 1872. The earthquake was felt as far away as Salt Lake City; it damaged buildings as far away as Chico.

The lines on the map are lines of equal intensity, which is a measurement of earthquakes based on damage. The center, or bull's eye, contains the hardest hit area. The area marked "IX" was the next most severely damaged.

The Roman numerals designating the zones are based on the modified Mercalli Intensity Scale (see fig. 52). Intensity, a measure based on subjective experience and damage, differs from "magnitude," a measure of the power of an earthquake based directly upon mathematical measurements.

Neither this scale of intensity nor scales of magnitude were developed in 1872. This map was drawn from historical accounts of the earthquake. Although it was prepared in 1972, in commemoration of the centennial of the 'quake, it was the first such map published for that earthquake.

Using this kind of information, the late Francois Matthes was able to construct in drawings his interpretation of what Yosemite Valley looked like at various stages. Two of his drawings are shown on pages 137-138. They give us a very clear idea of what one man saw on looking carefully and imaginatively into the past. The facts on which he based this picture are clear, but, like everything else in science, can be interpreted in other ways. Recent work, equally careful and imaginative, has given us a new picture of the few million years just before our time. According to this later view, few, if any, of the peneplains existed. There are similar levels to be sure; Dr. Matthes was correct in that. But the similar levels are to be accounted for not by the wearing down of entire mountain ranges to sea level, but by the weathering out of giant steps.

This later interpretation indicates that, in areas of granitic rocks, a giant staircase is produced within the mountain range by the normal processes of erosion. Where granitic rock is buried, it commences to turn to soil much more quickly than where it is not; the ground water, working on the rocks, has much longer to break them down than it does where the water is shed quickly from the rock surface. Once weathering starts, it enlarges the flat area, where still water can work, while the steeper parts, where rushing water carries the loose material away, remains steep.

Calling it a staircase implies an evenly graded flight, which is not true at all. The stairs are surrealistic, with uneven beginnings and strange continuations. Most of the steps face the San Joaquin Valley at a variety of levels, but some lead upward from steep canyons in other directions.

What is happening to the Sierra today? Are the everlasting hills truly everlasting? Even a brief trip into the high mountains or a drive through the gold country will give clear witness to eternal change – even in so enduring an element of creation as the bulky Sierra. Earthquakes, rock falls, rock slides, landslides – these are all fast means of changing the landscape. Anyone who has hiked the high mountains is well aware of falling rock, and of the danger of rock slides. All of these, from the single pebble cascading from a height to landslides involving entire mountains (such as Slide Mountain, which slid in 1842), are attempting to erode the peaks to level ground. In winter, snow avalanches speed down the rock ribs of the peaks. In the course of years, some of them have worn channels 50 to 100 feet deep, abraded smooth by the sliding rock, snow, and ice. The areas around Mt. Whitney and Sequoia and Kings Canyon National Parks are rich in these strange landforms, which resemble giant children's slides. A particularly good set can be seen across the canyon from Bearpaw Meadow on the High Sierra trail.

A process of change that takes longer, but is still within our immediate ken, is that of running water. We can see sandbars shift from year to year; we can watch gravel bars build up in a season; we can see sand, gravel, and rock being whirled along by a mountain stream or a sudden flash flood. It takes longer to observe some of the slower processes, but it is nonetheless possible. The birth and slow death of Kern Lake was nearly within one lifespan. Created in 1867 by a dam formed by a landslide, the lake had built a sandbar by 1916, which by 1928 had enlarged to cross the lower end of the lake. Gradually, the lake is being filled; eventually it will be a forest.

The most notable processes that are not instantaneous, but nonetheless observable, are the changes man himself makes. In a little more than 100 years, a network of roads has crossed and recrossed the Sierra. Sometimes it seems that there is little space left that is not roadway, yet it takes the concerted effort of all those who would preserve some vestige of the high country to keep even a small wilderness inviolate.

More striking than roads are the towns and would-be towns; the acres of wilderness carved into "developments" where erosion can race unchecked but where no people live. Mining and lumbering once threatened to destroy the very mountains that gave them existence; today, public pressure and more concerned industries have checked some of the excesses that destroyed wildlife, flooded towns, gashed the mountains, and left trash for beauty.

The threat today to the wilderness comes as much from its more thoughtless supporters as from those who wish to "win" it to civilization. The changes for "recreation" give us litter on even the mountain crests nearest heaven, parking lots where mountain meadows were, ephemeral summer cities amongst the tallest forests.

It is true that nature is more violent than man; that natural processes are more effective than human ones alone, but it in no wise excuses us from responsibility or culpability. Those of us who value beauty are glad that nature still has the upper hand, but it behooves us to watch the hand of man.

There are still slower natural processes that we can measure and sometimes observe. It may be that many human changes in the mountains are of this slow-acting sort: we will not know all the consequences of our actions for many generations. There are innumerable worldwide examples of such human alterations: Greece, once forested, is now a nearly arid land; the cedars of Lebanon, that furnished lumber for millenia, are almost gone. In our own land, we have murdered animals to the last survivor: the last eskimo curlew is gone; the passenger pigeon no longer lives; the grizzly, symbol of California, does not roam the Sierra; many other animals and plants we have driven, by our encroaching civilization or by our cruelty, to the brink of extinction. They will be fossils only, and in our day.

We have started other natural processes of which we cannot – or will not – foresee the consequences. We can measure some of nature's own slow processes, such as the movement of a glacier, the development of an alluvial fan, or the filling of a reservoir; but we have far too little knowledge of the long-term effects of our own actions. Scraping the surface of one of the granite "steps" for development may allow accelerated erosion to remove all soil, so as to give us bare rock where once a forest grew. Thoughtless forestry or mining practices may allow streams to carry sediment to the sea at a rapid rate, in that way removing the soil that supports the forest, and, in turn, destroying the watershed and causing floods. By altering the cloud cover, we may bring on another ice age. By our actions we can destroy mountains, poison water, eliminate life, or change our land to desert or ice field.

Chapter 11

"THE MOUNTAINS GROW, UNNOTICED"

Science is modern mythology, in that it seeks to explain the "why" of the world. To do so, we must take into account what facts we have, and build on them. It is not enough merely to consider the facts of Sierran history as shown in these pages – although that must be done. We cannot divorce the Sierra from the world; its story must fit with the story of the whole earth.

No geological "explanation" (or myth, if you will) has captured the American fancy like the current theory of "global tectonics" of which "continental drift" is a part. It is not a new idea; many people have looked at a globe of the earth, and, noting the close fit of Africa and South America, have wondered if perhaps they once were joined.

Global tectonics proposes just that: that once the continents were joined, and have since been torn apart. It is a unifying theory that not only reunites continents, but explains how mountains and valleys are formed, what is happening to the seas, and why land and water are distributed as they are. To be able to treat the earth as a unit (which it is) is satisfying to earth scientists, and fires the imagination of non-scientists. Derived from the Greek word for "builder," tectonics, particularly global tectonics, tells how the earth is built.

It may seem incredible, at first, to think of continents either "drifting" or being ripped apart, but remember that time is the friend of geology, and that what seems impossible in our limited experience may happen easily in the long reach of time. The solid earth may lose its solidity when viewed through time. What seems to us tough and immutable may be seen as fluid and mobile under the mitigating influence of time. In fact, the solid earth seems not so solid now when we delve beneath its skin.

We know very little about the world beneath our feet. The deepest hole we have managed to dig (for oil) is but six miles deep – scarcely a pin prick, in comparison to the 4000 miles to the center of the planet. Man has gone underground in person to see the rocks down only two miles, and then in a hot, dark South African gold mine. What little we know – or think we know – comes to us, not by direct observation, but by courtesy of earthquakes.

By measuring earthquake waves as they pass through the earth, scientists have derived an idea of what is beneath us – the earth we cannot see (figure 54). The innermost core apparently is solid, surrounded in turn by a transition layer, then a liquid outer core. This, in turn, is enclosed by the mantle, upon which the crust – that part of the earth on which we live – rests.

The crust of the earth is not uniform. It varies in thickness from 3 to 35 miles – far thinner, proportionately, than the shell of an egg. It is thinner under the ocean than under the land. Mountain ranges within the crust stand higher than their surroundings; here, the crust has an obvious upward lump. Some mountain ranges, including the Sierra Nevada, consist of rocks of lighter weight than other rocks in the earth's crust, or than the rock in the subcrustal part. Such mountains seem to have a "root" – a downward projection of lighter material into the heavier rock below, much like an iceberg in water. The portion we see is but a small part of the entire mountain mass, when its root into the crust is considered.

Beneath the crust, which contains the sedimentary rocks of the world, as well as volcanoes and metamorphic zones, is the "mantle," an 1800-mile-thick zone that contains rock that is more plastic than rocks at the surface. The outer part of this zone may be the mother of most of the surface rocks; it may be from here that volcanoes take their source, and from which the cores of mountains are ultimately derived.

No one has seen any rock that definitely has come directly from the mantle. Most of the evidence of the mantle's existence comes from earthquake waves. Yet because it may be the direct ancestor of our surface rocks, scientists have a great interest in it. In the 1950s, a project was undertaken to drill a hole in the earth deep enough to reach the mantle so that we could see, firsthand, what this rock looks like. Designated the "Mohole" project, it was successful in drilling a few preparatory holes, but none deep

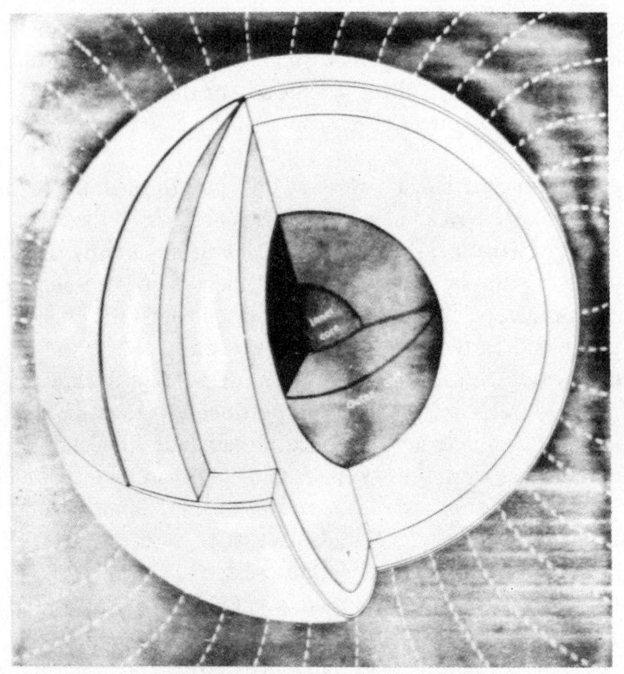

Fig. 54. Interior of the earth, as deduced from evidence provided by earthquakes. The center of the earth is about 4000 miles beneath our feet. It lies in what seismologists call the earth's "core," which comprises about 15 percent of the earth's volume. The core is separated, on the basis of the manner in which earthquake waves behave when they encounter it, into an inner kernel, believed to be solid, and a liquid outer core, with a transition zone between. It has been estimated that the core of the earth is over 80 percent iron.

Above the core is the earth's mantle, which contains about 84 percent of the volume of the earth. Between is another transition zone, which shows evidence of being bumpy, rather than smooth, as this drawing indicates.

Indeed, it is probable that some of these zones, such as the boundary between the crust and mantle, are far less regular than indicated here. We know by direct observation only the outer skin, or crust, of the earth, which constitutes only 1 percent of its volume. To us, it has immense topographic irregularities, yet projected to this scale, they disappear. Not all zones, however, are bumpy; seismic evidence from earthquake waves indicates that the boundary of the core itself is rather smooth.

enough to reach beyond the crust. Since the earth's crust is much thinner (3-5 miles) at the bottom of the sea than on land, where it is 12 to 35 miles thick, the principal, crust-penetrating hole was to have been drilled in the sea.

The undertaking would have been expensive and difficult, and did not get well started, either by Americans or by Russians, both of whom were trying. There is doubtless a great deal to be learned from an actual sample of rock from the earth's mantle, but today many scientists put a higher value on other earth projects. Meanwhile, most of us have already seen rocks from the moon!

Using what information we have about the interior of the earth and new findings about the depths of the sea, scientists have devised the theory of global tectonics to explain the cause of mountains and valleys, of continents and seas. According to this theory, the upper part of the earth is divided into several rigid segments or "plates," each plate consisting of a continent of a continent and adjoining seas. The plates, which make up the crust of the earth, are able to move about on top of the earth's more plastic mantle (figure 55).

At one time – perhaps 200 million years ago – the continents were joined together as one mass (figure 56), but have since been pulled apart as the plates on which they were riding moved. Evidence that this is so rests on the neat match of the margins of some continents (particularly if the lip of the continent, or continental shelf, and part of the continental slope are included); the similarity of plants, animals, fossils, and rocks in continents now separated by the sea; drastic changes in climate as shown by tropical fossils near today's poles and glacial deposits in today's tropics; and by the discovery of magnetic stripes in the ocean floor.

This last discovery has been particularly revealing. The centers of most of the world's oceans are marked by long ridges. In the Atlantic, the mid-Atlantic Ridge nearly bisects it; similar ridges mark the Pacific (the "East Pacific Rise," the "Pacific-Antarctic Ridge"), the Indian, and other ocean basins, forming a world-encircling ridge system. The ridges are unusual in their length, and have a deep cleft, in some places thousands of feet deep, that

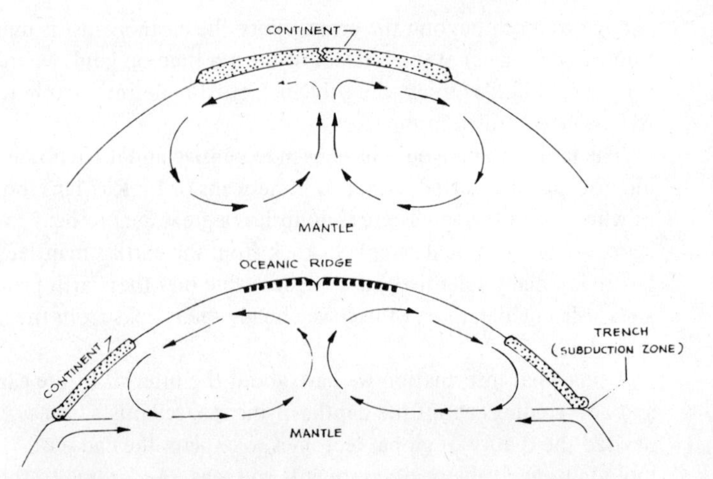

Fig. 55. How continents might break apart and the sea floor "spread." In the top diagram, a continental mass has found its way over a pair of heat convection "cells." As the heat slowly rises, the continent is ripped apart, each half drifting away.

In the lower drawing, heat from convection cells is shown rising along an oceanic ridge. Molten rock is brought up to pour out along the top of the ridge, forming stripes on either side, thereby widening the sea floor and pushing the continents farther apart.

The continents move apart until they meet an obstacle. As shown here, they encounter a deep trench, where currents are flowing into the earth's mantle. Here older rock is being pulled into the trench and ultimately into the mantle to be remelted, eventually, perhaps, to rise as molten rock along a ridge.

Convection currents were among the first "engines" suggested as mechanisms that might drive the continents apart. For many years, continental "drift," or movement apart, was doubted, partly because there seemed to be no force capable of rending continents asunder. Today, there is ample evidence that continents have moved apart, but scientists are still not in accord concerning the driving mechanism. It is a subject that is hotly debated among earth scientists.

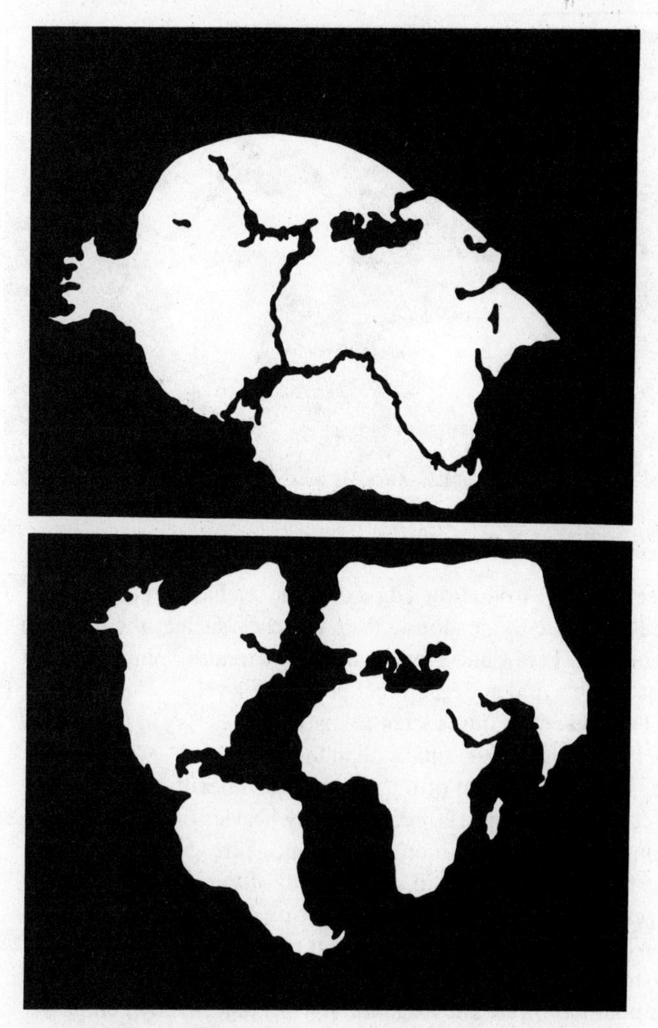

Fig. 56. *Top:* The earth's continents, jointed into one, as they may have been in the Paleozoic Era. *Below:* The continents as they began to drift apart in the Mesozoic Era.

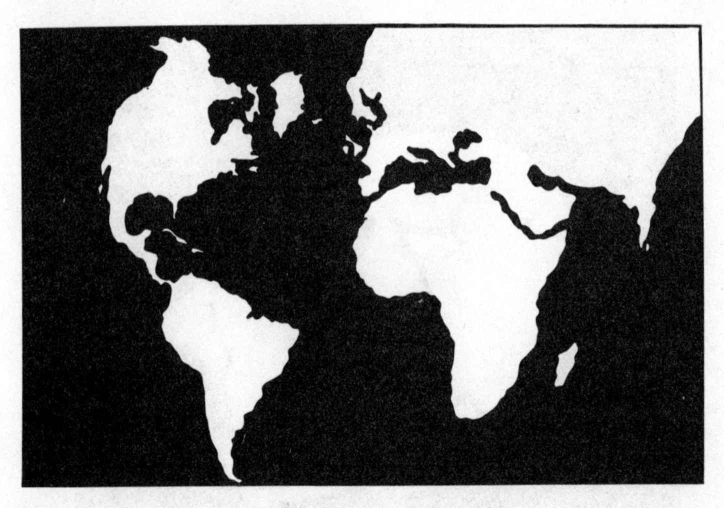

Fig. 57. The continents today.

bisects them. Around the edges of some sea basins, particularly the Pacific, are tremendously deep trenches, such as the Aleutian Trench, the Peru-Chile Trench, and the Marianas Trench, deepest spot in the world at 36,000 feet below sea level.

The mid-ocean ridges seem to mark "hot spots" in the earth, where heat rises more rapidly than in other places. It has long been known that most of the ocean floor is basalt, a fine-grained volcanic rock (see p. 120) considerably heavier than granite or the sedimentary rocks common on the continents. Now we know that fluid basalt rises up in or pours out of the center of the ridges periodically, to form lava flows on either side.

While the basalt was still liquid, it contained tiny floating crystals of magnetic minerals, especially magnetite. These crystals pointed toward the magnetic (north) pole, as any compass would. When the lava hardened into rock, the tiny mineral compasses were frozen into the position they had assumed, pointing toward the magnetic pole. When next the lava poured out or welled up, it took its place in the center of the spreading ridge. Gradually, older stripes migrated to the edges of the basin as more and more stripes were added to the center. In this way, the basalt stripes became lined up across the ocean floor, youngest in the center, oldest at the edges.

Scientists were able to discover that there were such things as basalt stripes because of the tiny compass needles oriented toward the magnetic pole. After all, the ocean is deep and dark; men have seen little of the bottom, and the samples they have taken with their meagre instruments (rather like a tablespoon on a mile-long string) are neither large enough nor numerous enough to give much of a picture of the deep sea floor.

But recording instruments, measuring magnetic properties, told a surprising story. Each new stripe on either side of the ridge was oriented opposite to the older one. If the new one pointed toward today's north as the magnetic pole, the next older ones on either side of the ridge pointed south! The one next to that north again, then south, then north, and so on. Why this should be – why the poles should flip flop – is not yet clear, but that magnetic "north" is changing end for end is clear.

As each new stripe found room for itself, something must have happened at the edges of the basin. Scientists now think that this is where the action is: here the oldest stripes of basalt dive into the deep trenches under the continental masses, wrinkling the continents into mountains, causing earthquakes and exploding volcanoes. Eventually the plate goes deep enough into the earth to be reunited with the mantle; to be remelted; eventually to be resurrected again as new rock from the ocean ridges.

This is how mountains form: as the old plate edge is pushed beneath the continental mass, the edge of the continent and the edge of the sea are bent and broken into mountains, or as one plate collides with another, mountains are born of the crash.

About 150 million years ago, the area that is now the Sierra Nevada must have been at the edge of the continent, under shallow seas. Offshore, volcanoes formed an arc of islands around it, similar to today's Aleutians. The region was part of the leading edge of one of the tectonic plates, and was gradually drifting northwestward (figure 58).

Measurements show that most rock debris that is eroded from the continents by rain and rivers, wind and ice eventually comes to rest on the continental shelf and its seaward slope – only

Fig. 58. How the Sierra might have formed.

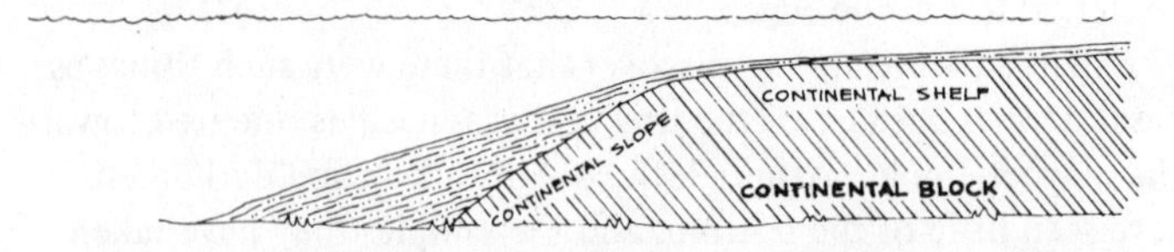

A. The area that is now the Sierra Nevada lies at the edge of the land. The continental block was its foundation; sediments were deposited on the gently sloping continental shelf, and a thick wedge formed on the steep continental slope. Beyond, out of view, lies the deep sea.

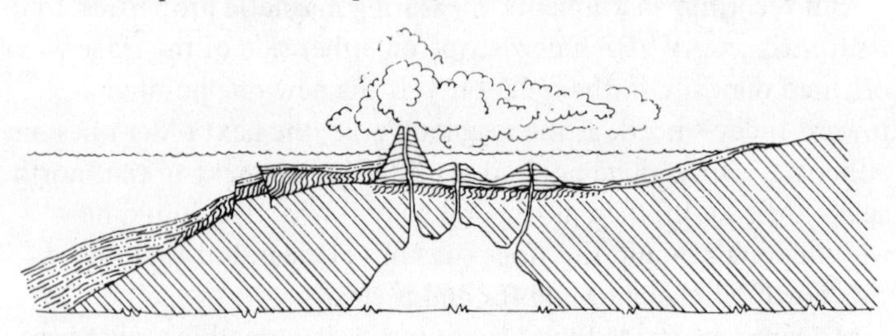

B. A chain of offshore volcanoes, some undersea, developed. Pressure was exerted from the west (left), perhaps because a "subduction zone" – a trench in which torn-off parts of the earth's crust are plunging downward into the earth's mantle – was adjacent. Pressure, breakage of rock along faults, together with heat, have metamorphosed the sedimentary rock.

about 10 percent makes its way into deeper water. Since fossils and other evidence indicate that the sediments that later became the metamorphic framework of the Sierra Nevada were deposited in fairly shallow water, scientists assume that the sea shelf and basin kept sinking, for a great thickness of rock accumulated there. Indeed, this seems to be a general rule for long, narrow sea troughs, such as the one that was where the Sierra is now: as the sediment accumulates, the basin sinks to accommodate it. In this way, the continents are growing outward by adding layers of rock fragments to their shores.

But the North American continent could not continue expanding at the expense of the deep ocean. This was the edge of the continent, where things happen. As one of the tectonic plates plunged downward toward the mantle, sediments – now rock, in

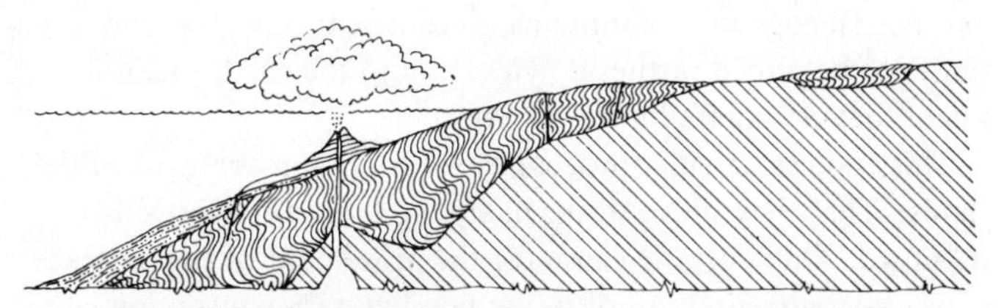

C. Most of the sedimentary rock has been transformed into metamorphic rock by heat and pressure; the area now rises gently above sea level toward the east, while parts of the west are still covered by water.

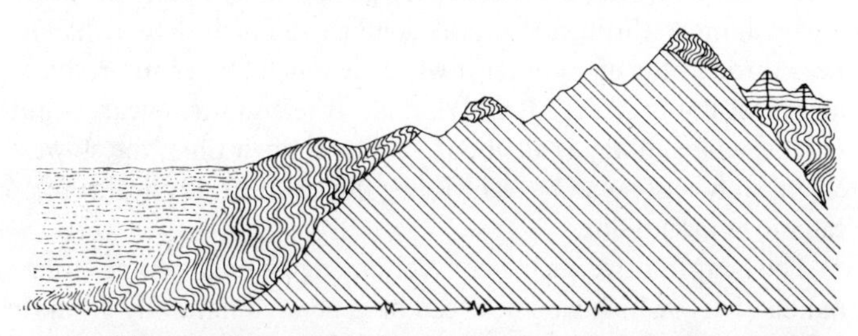

D. Metamorphism of the old rocks has been completed. Part of the underlying granitic rock has been exposed and eroded; the range has been uplifted; faults now mark the steep east face. To the west, what is now the Great Valley has been filled by sediment, covering much of the metamorphic rock. Not to scale.

the deeper parts of the basin – once on the ancient sea shelves, were buckled, bent upward and downward as they were squeezed and metamorphosed in the crustal nutcracker.

When the Sierra was so squeezed – if this is its true story – parts of the sedimentary rock were dragged downward far enough to get hot enough to melt. This melted rock, perhaps together with fluid rock from the earth's mantle and melting parts of an adjacent, downward-plunging crustal plate, became the magma that consolidated into granitic rock – the Sierran core.

So the land builds outward toward the sea, and is raised from the waters into mountains. The sea is widened by adding new rock from the depths of the earth, and by pushing against the continents and forcing them back. As the sea is widened in the center, its edges plunge into deep trenches to lift and crumple the continental edge into mountains, to cause earthquakes, and, eventually, to be pulled farther downward into the earth's mantle where they are melted into new rock.

This endless, circular process depends upon movement of the plates. That they move seems likely, but why they move is unknown. One possible cause for the movement of the plates is convection currents. All of us are familiar with convection currents in air – the heat rising from a stove, for example; but few have stopped to imagine similar stirrings within the solid earth itself. Such stirrings are most likely produced by heat. Obviously, heat will move through the more solid earth much slower than it rises through the air, so that it will take much longer to produce any measurable effect. It will, in fact, take countless years – but time is what geology is all about. Given enough time, the slowest chemical reaction can be significant; the slightest physical force can move mountains.

The heat to drive such a convection engine might come from radioactivity, which we know can be generated naturally from the minerals and rocks within the earth, or from some other unknown source.

Some geologists have proposed that tides may be powerful enough to start and maintain such earth motion. Not only water tides, which are quickly responsive to astronomical events, but earth tides, building up stress in the soft part of the mantle, pull and push the continents. The stress may be relieved as a sudden fault break, causing an earthquake; or it may be transformed into heat, rising and circulating.

In the past few years, the problem of what to do with the nuclear waste of the world has been plaguing scientists and politicians alike. One proposal is to throw it into one of the deep trenches, where one crustal plate is being forced under another. Since no one has actually seen the plates' actions ("the mountains grow, unnoticed"), it seems most unwise to dump such dangerous waste where it cannot be closely watched. It might well be returned to us in an incalculably destructive way.

What the future holds for the earth or the Sierra Nevada we do not know. Will we poison the earth with nuclear wastes or in some other way so that no life can exist? Are we entering, or, by our actions, producing a new glacial stage? Can we see, in the Sierra, forerunners of new volcanic activity, with perhaps new granite forming deep within the earth?

Whatever the future holds, the years of our individual lives are but a flicker in the long days of earth, and even the earth's eventful history is short measured against the timeless immensity of the universe. We are a middling planet in a small solar system in a minor galaxy. Yet time is all we have.

Appendix

HOW TO USE THIS BOOK

This book emphasizes processes. There are some specialized words in it, but most of them are defined in tables, in the rock key, or in the appendix. Try to get a sense of what has happened to the mountains. You may decide that no one knows, and, in a way, you will be correct. All that can be done for you is to give you what other students of the Sierra think or have thought. None of them is "right" or "wrong." Ideas must be revised as fast as new facts are discovered. This is the scientific method, and the way of all learning.

To get the most from this book, and certainly to enhance your enjoyment, go into the mountains with your eyes opened to geology. If you do not already have a good idea of the different kinds of rocks, try to identify them by using the key beginning on page 203. You will make some mistakes; no key is foolproof, and rocks are difficult. After you have decided what you think the rock may be, check it in tables 2, 3, 5 and 6. If it seems to fit the description, and to be in the right place, perhaps you have correctly identified it.

Your answer, derived from the key, will be a fairly simple rock name. It is possible – and likely – that any geologist you ask will give it a more complex name. That does not matter; if you have identified it within a broad group, you will have taken the first step in understanding the process by which it got there and what it represents. Refinements in naming can come later.

In fact, the whole problem of naming is quite complex, for a variety of reasons, some historical and some technical. It is difficult for the ordinary person unversed in the terminology to read geologic reports simply because the words are so strange. To make it easier for you to read geologic reports on the Sierra Nevada, the glossary beginning on page 193 lists most of the technical names that have been given to rocks of the Sierra. It does not list the so-called "formation" name, which may have little to do with the type of rock or its origin. By referring to this list, you can discover what sort of rock is meant by the names you encounter, even though you may not understand all of the fine distinctions as yet.

If the key does not give you enough help, try to visit one of the localities listed in tables 2, 3, 5 and 6. Although these are good examples of the different kinds of rocks, they are merely samples, not by any means an exhaustive list of all of the outcrops! Other tables in the book will point out places to see volcanic, glacial, and other features of the Sierra.

This book is, of course, just a beginning. To list all of the geologists who have worked in the Sierra Nevada this past century and a half – and whose work was the source of this book – would take more than two pages; a list of every article that has been written on the geology of the Sierra Nevada would run to thousands of entries. On this page is a starter kit of information on the geology of the Sierra Nevada. With the exception of books by John Muir and some by the late Francois Matthes, the works are highly technical and may require assiduous use of the dictionary to understand the many specialized words.

The words may be difficult, but the ideas are stimulating. Bear this in mind: after all of the "formations" are named and renamed; after the processes are described and redescribed; after geologic theories are thought and rethought; after all of the technical terms are defined and redefined, the mountains still abide. The rocks are there for you to study; you can find new facts and develop new theories, or you can deepen your appreciation of the mountains by understanding a bit more of the story told by them.

SUGGESTIONS FOR FURTHER READING

Geology of northern California. Edited by Edgar H. Bailey. Issued as Bulletin 190 by the California Division of Mines and Geology, P. O. Box 2980, Sacramento, California. 1966.

"Geology of the Sierra Nevada," by Paul C. Bateman and Clyde Wahrhaftig, is the best summary available of the geology of the Sierra. Both authors are experts in the range, having worked many years there. Also included in the bulletin are papers entitled "Geology of the Taylorsville area, northern Sierra Nevada," by Vernon E. McMath; "Tertiary and Quaternary geology of the northern Sierra Nevada," by Cordell Durrell; "Cenozoic volcanism of the central Sierra Nevada, California," by David B. Slemmons; and "Economic mineral deposits of the Sierra Nevada," by William B. Clark.

Geologic map of California. Compiled under the direction of Charles W. Jennings. Published by the California Division of Mines and Geology, P. O. Box 2980, Sacramento, California.

These sheets cover the Sierra Nevada: Westwood (Susanville), Sacramento, Walker Lake, San Jose, Mariposa, Fresno, Bakersfield, and tiny parts of Death Valley, Los Angeles, and Trona.

California's changing landscapes. By Gordon B. Oakeshott. Published by McGraw-Hill Book Company, San Francisco and New York. 388 pages. Issued both in paper and cloth. 1971.

A summary of the geology of the whole state, with two chapters on the Sierra Nevada. Less technically written than the two listed above.

John Muir's studies in the Sierra. Edited by William E. Colby. Published by the Sierra Club, San Francisco. 103 pages. 1960.

The mountains of California. By John Muir. Doubleday & Company, Garden City, New York. 300 pages. 1961.

Two of John Muir's many works on the Sierra Nevada. *Studies* is a republication of his articles in the *Overland Monthly* in 1874 and 1875; *Mountains,* published in 1894, was the first of his work to be published as a book. Lewis Osborne, Ashland, Oregon, has issued other Muir books, as have other publishers.

Sequoia National Park, a geological album. By François Matthes. Published by the University of California Press, Berkeley. 136 pages. 1950. Edited by F. Fryxell.

The incomparable valley. By François Matthes. Published by the University of California Press, Berkeley. 160 pages. 1950.

Two of Dr. Matthes' less technically written books that are a must for any lover of the Sierra.

ROCK NAMES OF THE SIERRA NEVADA

The following rock names are those you will encounter in reading technical reports on the geology of the Sierra Nevada.

The definitions given here are simplified from those in the *Glossary of geology,* published in 1972 by the American Geological Institute. Names in italics are those used in this book; they are "field terms" suitable for identification of rocks without the aid of a microscope or chemical laboratory. Other names, not set in italics, are referred to these simpler terms so that you will have a general understanding of their meaning.

Adamellite (igneous – plutonic). Synonym for quartz monzonite, a member of the *granite* family. (See *granite.*)

Agglomerate. Mixture of coarse, angular, volcanic rock fragments.

Agmatite (igneous and metamorphic). Mixture of broken but reconstituted igneous and metamorphic rock fragments.

Alaskite (igneous – plutonic). Light-colored variety of *granite.* (See *granite.*)

Albitite (igneous – plutonic). Type of *porphyry* in which both the large crystals and fine-grained background are of the albite variety of feldspar. (See *porphyry.*)

Amphibolite (metamorphic). Type of *gneiss* or *schist* with dark (amphibole) minerals and feldspar, but little or no quartz. (See *gneiss* and *schist.*)

Andesite (igneous – volcanic). Fine-grained, gray to dark-colored rock that has the same chemical composition as *diorite.* (See p. 119.) Sometimes another rock name is added to indicate texture: andesite porphyry has the same mineral composition as *andesite,* with the texture of *porphyry*; andesite tuff is *tuff* (volcanic *ash*) that has the chemical composition of *andesite.*

Aplite (igneous – plutonic). Light-colored, sugary-textured rock.

Arenite (sedimentary). *Sandstone.* (See *sandstone.*)

Argillite (sedimentary). Hard variety of claystone. (See claystone.)

Arkose (sedimentary). Coarse-grained *sandstone,* commonly pink, composed of grains of feldspar and quartz. (See *sandstone.*)

Ash (igneous – volcanic). Very fine particles blown from a volcano. Volcanic "ash" is not the remains of the burning of coal or wood, but is made up of tiny fragments of red- or white-hot natural glass or rock. (See p. 120.)

Basalt (igneous – usually volcanic). Dark, fine-grained rock that has the same chemical composition as *gabbro.* (See p. 120.)

Basement (igneous or metamorphic). Rocks that are beneath and older than the oldest unmetamorphosed rocks in an area.

Bastite. Variety of the mineral *serpentine.* (See *serpentine.*)

Bedrock (igneous, sedimentary, or metamorphic). Rock that underlies soil or unconsolidated material.

Breccia (igneous, sedimentary, or metamorphic). Rock made of sharp-edged fragments, large and small. Consolidated rubble. Sometimes another rock name is attached to indicate composition: andesite breccia is *breccia* made of *andesite* fragments.

Calcarenite (sedimentary). *Calcareous rock* made mostly of grains of calcareous sand. (See *calcareous rock.*)

Calcareous rock (metamorphic). Term may be applied to either sedimentary or metamorphic rock. Virtually all *calcareous rocks* in the Sierra Nevada have been metamorphosed. *Calcareous rock* includes limestone, marble, dolomite, calcarenite – all composed of an appreciable amount of calcium carbonate ($CaCO_3$). Carbonate rock technically has more $CaCO_3$ than *calcareous rock,* but in this book both carbonate and *calcareous rocks* are called *calcareous rock.* Both will "fizz" when acid is placed on them. (See p. 204.)

Calc-hornfels (metamorphic). *Hornfels* whose chemical composition includes considerable calcium carbonate ($CaCO_3$). Variety of *calcareous rock.* (See *hornfels* and *calcareous rock.*)

Calc-silicate (metamorphic). Variety of *calcareous rock* containing a mixture of calcium carbonate ($CaCO_3$) and silicate minerals rich in calcium. (See *calcareous rock.*)

Camptonite (igneous – plutonic). Variety of *diorite.* (See *diorite.*)

Carbonate rock (sedimentary or metamorphic). Rock consisting chiefly of calcium carbonate ($CaCO_3$). (See also *calcareous rock.*)

Chert (sedimentary). Very hard, dense, tough rock consisting principally of silica (SiO_2). Color varies, but Sierran *chert* is commonly in thin red, green, or yellow beds. Usually has conchoidal fracture. Same as flint. (See p. 62.)

Cinder (igneous – volcanic). Fragment of glassy or fine-grained volcanic rock sufficiently fine-grained that its nature cannot be determined by the naked eye. It is vesicular (has many holes). (See p. 120.)

Clay (sedimentary). Very fine-grained sedimentary material easily moldable when wet (plastic). Also the name for a family of minerals. (See p. 99.)

Claystone (sedimentary). Very fine-grained rock made from *clay*. (See *clay*.)

Coal (metamorphic). Black rock, made of carbon, that will burn. Sierran coal is very soft. (See p. 99.)

Conglomerate (sedimentary). Rock consisting of sand-sized fragments, together with dominating, larger rounded rock or mineral fragments. (See p. 98.)

Dacite (igneous – volcanic). Variety of *andesite* having more quartz and less feldspar than most andesite. (See *andesite*.)

Diabase (igneous – plutonic). Variety of *diorite* consisting of plagioclase feldspar (labradorite) and dark mineral (pyroxene). (See *diorite*.) Crystals of feldspar are included within the pyroxene.

Diatomite (sedimentary). Variety of *shale* composed largely of the remains of diatoms (microscopic plants). Common in California, but not in the Sierra Nevada. (See *shale*.)

Diorite (igneous – plutonic). Usually gray; composed mostly of plagioclase feldspar with less than half dark minerals and almost no quartz. (See p. 78.)

Dolerite (igneous – volcanic). Synonym for diabase. (See diabase.)

Dolomite (sedimentary). *Calcareous rock* consisting largely of the mineral dolomite ($CaMg(CO_3)_2$). (See *calcareous rock*.)

Dunite (igneous – plutonic). Variety of *peridotite* consisting almost entirely of the mineral olivine. (See *peridotite*.)

Epidote-garnet rock (metamorphic). Rock consisting mainly of the minerals epidote and garnet. (See *schist* or *gneiss*.)

Fanglomerate (sedimentary). Coarse-grained rock, similar to *conglomerate* except that the pebbles are more angular. Originally deposited in an alluvial fan. (See *conglomerate*.)

Flows, lava. See *lava*.

Gabbro (igneous – plutonic). Coarse-grained, dark-colored rock consisting principally of plagioclase feldspar and more than half dark minerals, without quartz. (See p. 78.)

Gneiss (metamorphic). Coarse-grained, banded rock, in which layers of granular minerals alternate with layers of flaky ones. (See p. 62.)

Granite (igneous – plutonic). Rock consisting of visible mineral grains, chiefly quartz and feldspar, with some dark minerals. Usually has a salt-and-pepper look. (See p. 78.)

Granodiorite (igneous – plutonic). Member of the *granite* family. Granodiorite contains more plagioclase feldspar than true *granite* and less than quartz diorite. (See *granite.*)

Granophyre (igneous – volcanic). A variety of *porphyry*. (See *porphyry.*)

Grauwacke. Alternate spelling of graywacke.

Gravel (sedimentary). Unconsolidated accumulation of larger rock particles and sand. (See *conglomerate,* p. 98.)

Gravestone schist (metamorphic). Type of *schist* (see *schist*) that commonly protrudes from the ground in Sierran foothills; from a distance, outcrops may resemble tombstones.

Gravestone slate (metamorphic). Type of *slate* (see *slate*) that commonly protrudes from the ground in Sierran foothills; from a distance, outcrops may resemble tombstones.

Graywacke (sedimentary). Variety of *sandstone.* Usually dark colored, with an assortment of angular mineral and rock fragments. Common in California. (See *sandstone.*)

Greenstone (metamorphic). Dark, fine-grained rock which may be green owing to the presence of green minerals, but may also be brown, black, or reddish. Derived from basic or ultrabasic igneous rock, particularly basalt. (See p. 60.)

Grit (sedimentary). Variety of *sandstone* composed of angular particles. (See *sandstone.*)

Hornfels (metamorphic). Fine-grained rock in which the grains are not oriented in one direction. Grains may not be visible under the hand lens, but the rock will not split into thin layers, as do rocks with oriented minerals. (See p. 60.)

Ignimbrite (igneous – volcanic). Rock formed from consolidated volcanic *ash.* (See *ash.*)

Keratophyre (igneous). Variety of trachyte, grouped in this book under *andesite.* (See *andesite.*)

Lahar (igneous – volcanic). Mud flow originating from a volcano. (See p. 122.)

Lamprophyre (igneous – plutonic). Dark-colored *porphyry* (see *porphyry*) in which both the large crystals and fine-grained groundmass are dark minerals.

Lapilli (igneous – volcanic). Fragments of pea-sized *lava* blown from a volcano. (See *lava,* compare *cinder.*)

Latite (igneous – volcanic). *Porphyry* (see *porphyry*) in which the large crystals are feldspar. Rock has composition similar to *andesite.* (See *andesite.*)

Lava (igneous – volcanic). Rock cooled from molten state on or near the surface of the ground. (See p. 122.)

Leucotrondhjemite (igneous – plutonic). Member of the *granite* family. Contains quartz and plagioclase feldspar (oligoclase), with little dark mica (biotite). (See *granite.*)

Lherzolite (igneous – plutonic). Variety of *peridotite.* (See *peridotite.*)

Lignite (sedimentary). Soft coal, slightly metamorphosed. (See p. 99.)

Limestone (sedimentary). Fine-grained *calcareous rock* made principally of calcium carbonate ($CaCO_3$). Many limestone beds were derived from animal shells or reefs. Limestone in the Sierra Nevada has been metamorphosed. (See *calcareous rock.*)

Liparite (igneous – volcanic). Synonym for *rhyolite.* (See *rhyolite.*)

Marble (metamorphic). Crystalline limestone. (See *calcareous rock.*) Marble has been recrystallized, so that shiny crystals of calcium carbonate ($CaCO_3$) may show under the hand lens.

Mariposite (metamorphic). Bright green type of *schist* consisting largely of the green mica, mariposite. (See *schist.*)

Marl (sedimentary). *Shale* that is a mixture of clay and calcium carbonate ($CaCO_3$). Most marl in the Sierra Nevada has been metamorphosed. (See *shale.*)

Melaphyre (igneous). Dark-colored *porphyry.* (See *porphyry.*)

Meta-. Prefix indicating that rock has been metamorphosed. Rock names with this prefix can be found listed under the original name, i.e., meta-andesite (metamorphosed *andesite*), see *andesite;* metarhyolite (metamorphosed *rhyolite*), see *rhyolite.* The rock may not be exactly as described under its original name – it has been metamorphosed – but probably will be recognizable. The following "meta" rocks have appeared in reports on the Sierra Nevada: Meta-arkose; metabasalt; metabreccia; metachert; metaconglomerate; metadacite; metadiabase; metadiorite; metadolerite; metadolomite; metagabbro; metagranite; metagrauwacke; metagraywacke; meta-igneous; metakeratophyre; metalamprophyre; metalava; metalimestone; metamudstone; metamelaphyre; metaporphyry; metapyroclastic; metaquartz monzonite; metarhyolite; metasandstone; metasedimentary; metaserpentine; metashale; metasiltstone;

metatrap; metatuff; metatuff breccia; metavolcanic flows; metawacke.

Migmatite (igneous and metamorphic). Rock containing a mixture of igneous and metamorphic rock, often streaked.

Monzonite (igneous – plutonic). Member of the *granite* family containing plagioclase feldspar and orthoclase feldspar and dark minerals, but very little or no quartz. Commonly darker gray than *granite,* but not as dark as *diorite.* (See *granite* and *diorite.*)

Mudstone (sedimentary). Rock consolidated from mud. Does not break into thin layers as true *shale* does, but is otherwise similar. (See *shale.*)

Mylonite (metamorphic). Streaky, banded, fine-grained *breccia.* Its texture results from pulverization in a fault zone. (See *breccia.*)

Norite (igneous – plutonic). Dark-colored coarse-grained, close relative of *gabbro,* but has a different type of dark mineral (hypersthene). (See *gabbro.*)

Obsidian (igneous – volcanic). Dark-colored natural glass. (See p. 120.)

Opdalite (igneous – plutonic). Member of the *granite* family, related to *diorite.* (See *granite* and *diorite.*)

Orthoquartzite (sedimentary). Variety of *sandstone* (see *sandstone*). Orthoquartzite in the Sierra Nevada has been metamorphosed.

Peat (sedimentary). Unconsolidated deposit of decomposed plant remains; the initial stage in the development of coal. Peat is common in the Central Valley of California.

Pegmatite (igneous – plutonic). Very coarse grained. Can be any composition, but most *pegmatite* has the same mineral composition as *granite.* (See p. 79.)

Peridotite (igneous – plutonic). Coarse-grained, ultramafic rock, composed of dark minerals, with no quartz and little or no feldspar. (See p. 79.)

Perlite (igneous – volcanic). Natural glass, usually cracked into tiny round beads.

Phthanite (sedimentary). Old name for *chert.* (See *chert.*)

Phyllite (metamorphic). Silky rock that is layered. (See p. 61.)

Phyllonite (metamorphic). Type of *phyllite.* (See *phyllite.*)

Porphyrite (igneous). Same as *porphyry.* (See *porphyry.*)

Porphyry (igneous). Contains large crystals in a fine-grained groundmass (matrix). (See p. 79.)

Propylite (igneous – volcanic). *Andesite* that has been altered by hot water. (See *andesite.*)

Pumice (igneous – volcanic). Light-colored, glassy rock buoyant enough to float on water. (See p. 120.)

Pyroclastic rock (igneous – volcanic). Rock made of fragments blown from a volcano.

Pyroxenite (igneous – plutonic). Member of the *peridotite* family, consisting principally of the mineral pyroxene. (See *peridotite.*)

Quartz diorite (igneous – plutonic). Coarse-grained member of the *granite* family related to *diorite,* but containing quartz. (See *diorite* and *granite.*)

Quartz latite (igneous – volcanic). Medium-gray rock with a fine-grained groundmass and large crystals of quartz. A *porphyry.* Synonym of rhyodacite. (See *porphyry.*)

Quartz monzonite (igneous – plutonic). Member of *granite* family, with considerable feldspar (both orthoclase and plagioclase) and quartz, with some dark minerals. (See *granite.*) Rock with less plagioclase feldspar is called *granite;* with more, granodiorite.

Quartz veins. See quartz, p. 17.

Quartzite (metamorphic). Metamorphosed *sandstone* or *chert.* (See p. 63.)

Rhyodacite (igneous – volcanic). Variety of latite.

Rhyolite (igneous – volcanic). Fine-grained, light-colored member of *granite* family. (See p. 119.)

Rhyolite tuff (igneous – volcanic). Consolidated volcanic *ash* having the chemical composition of *rhyolite.* (See *ash.*)

Rödingite (igneous – plutonic). Variety of *gabbro.* (See *gabbro.*)

Sandstone (sedimentary). Medium-grained rock consisting of particles of rock or minerals cemented together by a natural glue (iron oxide, silica, calcium carbonate). (See p. 98.)

Saxonite (igneous – plutonic). Variety of *peridotite.* (See *peridotite.*)

Schist (metamorphic). Metamorphosed crystalline rock that splits readily into flakes or slabs. Most grains large enough to be seen easily. (See p. 61.)

Scoria (igneous – volcanic). Vesicular crust on the surface of lava flows. Heavier and darker than *pumice.* (See *pumice.*)

Serpentine (metamorphic). Used to designate both rock (also called serpentinite) and mineral (serpentine). Green to black rock, greasy appearing. May contain fibers of asbestos. (See p. 62.) Commonly from alteration of ultramafic igneous rock.

Serpentinite. See *serpentine.*

Shale (sedimentary). Fine-grained rock, derived from the compaction of clay, silt, or mud, characterized by its tendency to break into thin layers. (See p. 99.)

Siltstone (sedimentary). Rock derived from the compaction of dust-sized particles (silt). Similar to *shale.* (See *shale.*)

Skarn (metamorphic). Rock derived from the metamorphosis of *calcareous rock* to which has been added considerable silica ($Si0_2$).

Slate (metamorphic). Fine-grained rock that commonly will split into layers. Grains are too fine to be seen with a hand lens. Most *slate* has a shiny surface. (See p. 61.)

Soapstone (metamorphic). Soft rock with an "unctuous" feel, composed mostly of talc. May be *schist, phyllite,* or *slate* in texture. (See *schist, phyllite, slate.*)

Subjacent Series (obsolete). Refers to all of the Mesozoic and Paleozoic igneous and metamorphic rocks in the Sierra Nevada.

Superjacent Series (obsolete). Refers to all sedimentary and volcanic rocks deposited on top of the granitic and metamorphic rock (Subjacent Series) in the Sierra Nevada.

Syenite (igneous – plutonic). Member of the *granite* family. Contains orthoclase feldspar, some plagioclase feldspar, and dark minerals, but very little or no quartz. (See *granite.*)

Tactite (metamorphic). Metamorphosed (in a contact zone) and chemically changed *calcareous rock.* (See *calcareous rock;* also *skarn.*)

Tephra (igneous – volcanic). A general term for all fragments blown from a volcano.

Till (sedimentary). Mixture of boulders and fine fragments left by glaciers or glacier streams. (See p. 99.)

Tillite (sedimentary). Consolidated rock formed of *till.* (See *till.*)

Tombstone rock (metamorphic). Term used to refer to nearly vertical outcroppings of metamorphic rock in the foothills of the Sierra. From a distance, a group of outcroppings may resemble a cemetery. Also called gravestone slate, gravestone schist. (See p. 57.)

Tonalite (igneous – plutonic). Variety of *diorite* composed of plagioclase feldspar and dark mineral – usually hornblende – with considerable quartz. Same as quartz *diorite.* (See *diorite.*)

Trachyandesite (igneous – volcanic). Variety of *andesite.* (See *andesite.*)

Trachybasalt (igneous – volcanic). Variety of *basalt.* (See *basalt.*)

Trachyte (igneous – volcanic). Fine-grained rock resembling *rhyolite.* (See *rhyolite.*)

Trap (igneous – volcanic). Dark-colored, fine grained rock. Generally *basalt.* (See *basalt.*)

Troctolite (igneous – plutonic). Variety of *gabbro.* (See *gabbro.*)

Trondhjemite (igneous – plutonic). Member of *granite* family, similar to *granite,* except that it contains plagioclase feldspar and very little or no orthoclase feldspar. (See *granite.*)

Tufa (sedimentary). *Calcareous rock* formed around a hot spring. (Early reports sometimes used *tuff* and tufa interchangeably. Modern use restricts *tuff* to volcanic ash and tufa to hot spring deposits. Tufaceous refers to tufa; tuffaceous to *tuff.*)

Tuff (igneous – volcanic). Consolidated volcanic *ash.* (See *ash.*) *Tuff* may be mixed with clay, sand, or pebbles. Since it is usually deposited in layers (dropping from the air onto the ground or into water bodies), it is, in a sense, sedimentary. (See p. 120.)

Volcanic *ash.* See *ash* and *tuff.*

Wacke (sedimentary). Type of *sandstone.* (See *sandstone.*)

Wehrlite (igneous – plutonic). Variety of *peridotite.* (See *peridotite.*)

FIELD KEY TO ROCKS OF THE SIERRA NEVADA

Following is a key to rocks of the Sierra Nevada. It is not usable in other areas.

In all of the tests given, select a "fresh" piece of rock; that is, one that has not been exposed to weather. Most rocks become crusted after being exposed to air. Their color changes; their hardness changes; in fact, even the minerals themselves change. For that reason, break the rock open to see a clean, unweathered face.

If at all possible, try to obtain the rock "in place." Take it from an outcrop. That way, the sample will certainly be from the Sierra Nevada. Also, the outcrop itself – its location, shape, and aspect – are all clues to the history and identification of the rock.

Materials you will need in order to use this key are a hand lens or field magnifier, preferably 10 power or more; a small amount of vinegar or another acid; a pocket knife or nail; a prospector's geologic pick or other hammer; and a container of water. Protective glasses are not required for the identification of rocks, but are necessary to the safety of the investigator.

Rock will float on water . . . . See pumice, table 6

Rock will not float on water . . . . Go to 1

1. Rock will not float on water

   Rock consists almost wholly of quartz . . . . Go to 22

   Rock does not consist almost wholly of quartz or it is not possible to tell . . . . Go to 2

   One way to distinguish between the glassiness of quartz and the sheen of calcareous minerals is by scratching the rock or mineral with a nail or knife (see scratch test at 4). Other clues to help you identify quartz are given in table 1.

2. Rock does not consist almost wholly of quartz or it is not possible to tell

   Rock has individual pieces of rock or grains large enough to see by eye or with a hand lens . . . . . Go to 3

Very few or no individual pieces of rock or grains are
large enough to see by eye or with a hand lens . . . Go to 4

Texture of the rock – the size and arrangement of its mineral grains – is in question here. Color does not affect texture; fragments of rocks and minerals, whatever their color, are critical in determining texture. Very fine-grained rocks have grains too small to see either by eye or under a hand lens. Claystone is such a rock: too fine grained to show its individual minerals. For such a rock, choose "Go to 4."

If individual grains or fragments can be distinguished, either by eye or under the hand lens, the rock has coarse enough grain that you should select "Go to 3." Granite and sandstone are such rocks. The salt-and-pepper look of granite is due to individual mineral grains of different colors, which you can see if you look at the rock closely. Sandstone, observed carefully, may be seen to be made up of similar grains cemented together. Or it may have several types of grains cemented together. The gritty "feel" of sandstone is a clue that it is made up of individual particles, which can be seen by eye or under the hand lens. If you can, try to examine granite, sandstone, and clay or coal to see the difference in texture.

Beware also of "accidental" inclusions in a rock in which the grains are otherwise indistinguishable. If there are only one or two such inclusions (fossils, for example), you should nevertheless select the second alternative; if there are several, or numerous "chunks" in the rock (like raisins in pudding), choose "Go to 3."

3. Rock has individual pieces of rock or grains large enough
to see by eye or with a hand lens

Rock will effervesce ("fizz") if acid is placed on it . . See calcareous rock, table 2

Rock will not "fizz" if acid is placed on it . . . . Go to 8

If you use vinegar, a very weak acid, you may see "fizzing" only under your hand lens. Be certain that the bubbles – the "fizz" – are distributed more or less uniformly throughout the drop of acid.

Again: be sure to use a fresh surface for this test, as particles of dust may masquerade as "fizz."

4. Very few or no individual pieces of rock or grains are
large enough to see by eye or with a hand lens

Rock can be scratched by knife or nail on fresh
surfaces . . . . Go to 5

Rock cannot be scratched by knife or nail on fresh
surfaces . . . . Go to 6

Inspect the scratch carefully with your hand lens to be certain that the rock has been scratched. If you can see a metallic line on the rock, very likely the nail is not as hard as the rock. If there is a ridge of rock dust and an accompanying groove, the nail is harder.

5. Rock can be scratched by knife or nail on fresh surfaces

Rock will "fizz" if acid is placed on it . . . . See calcareous rock, table 2

Rock will not "fizz" if acid is placed on it . . . . Go to 26

See note at 3.

6. Rock cannot be scratched by knife or nail on fresh surfaces

Rock breaks in "scoops" (has conchoidal fracture) . . . Go to 57

Rock does not break in "scoops" . . . . Go to 7

Glass, when broken, shows conchoidal fracture on its edges. "Conchoidal" refers to the concentric, shell-like pattern developed on the broken part.

7. Rock does not break in "scoops"

Rock is wholly dark brown or black on fresh
surfaces . . . . Go to 10

Rock is not wholly dark brown or black on fresh
surfaces . . . . Go to 39

Beware of crystals or "accidental" inclusions. It is the over-all color that is critical.

8. Rock will not "fizz" if acid is placed on it

Rock has greasy look on fresh surfaces; may be
greenish or brown; may split along thin planes . . See serpentine, table 2

Rock does not have greasy look . . . . Go to 9

9. Rock does not have greasy look

Rock consists entirely of mineral grains or rock
fragments of approximately the same size . . . . . . Go to 45

Rock does not consist of mineral grains or rock
fragments of approximately the same size . . . . . . Go to 37

If part of the rock is too fine to see grains, yet the remainder consists of grains, rock fragments, or crystals large enough to see, choose "Go to 37."

If part of the rock has grains nearly all of one size, the remainder larger or smaller, choose "Go to 37."

If the grains are nearly all almost the same size (as are grains in granite, sandstone, and manufactured brick), select "Go to 45."

Again: beware of "accidental" inclusions. If there are one or two larger pieces, or a clot of color or smaller minerals, probably the first choice is correct.

10. Rock is wholly dark brown or black on fresh surfaces

Rock has shiny or satiny surface . . . . See slate, table 2

Rock does not have shiny or satiny surface . . . . Go to 11

11. Rock does not have shiny or satiny surface

Rock is vesicular (has holes) . . . . Go to 60

Rock is not vesicular . . . . Go to 12

12. Rock is not vesicular

Can see field exposure . . . . Go to 13

Cannot see field exposure . . . . Go to 14

13. Can see field exposure

Field exposures resemble columns . . . . Go to 15

Field exposures do not resemble columns . . . . Go to 16

14. Cannot see field exposure

Chip of rock when held to light has translucent edges . . . . See andesite, table 6

Chip of rock when held to light is opaque . . . . Compare basalt, table 6, and hornfels, table 2

15. Field exposures resemble columns

Chip of rock held to light has translucent edges . . . See andesite, table 6

Chip of rock held to light is opaque . . . . See basalt, table 6

16. Field exposures do not resemble columns

Field exposures on rolling hills with large boulders mixed with smaller material . . . . See andesite, table 6

Field exposures not on rolling hills with large boulders mixed with smaller material. . . . . Go to 17

17. Field exposures not on rolling hills with large boulders mixed with smaller material

Field exposures in layers . . . . Go to 18

Field exposures not in layers . . . . Go to 23

18. Field exposures in layers

Rock splits along bedding plane . . . . Go to 19

Rock does not split along bedding plane . . . . Go to 24

19. Rock splits along bedding plane

Rock breaks in "scoops" (has conchoidal fracture) . . . . See chert, table 2

Rock does not break in conchoidal pattern . . . . . . Go to 20

See note at 6.

20. Rock does not break in conchoidal pattern

Rock has "greasy" look on fresh surfaces; may be greenish or brown; may split along thin planes . . See serpentine, table 2

Rock does not have greasy look . . . . Go to 21

21. Rock does not have greasy look

Rock splits along parallel lines (bedding planes) to leave shiny smooth surface . . . . See slate, table 2

Rock splits along more or less parallel lines but does not leave shiny smooth surface . . . . See hornfels, table 2

22. Rock consists almost wholly of quartz

Rock may break in "scoops" (have conchoidal fracture) and has glassy look on fresh surfaces . . . See quartz, table 1

Rock may be in layers, may appear sugary . . . . Go to 33

See note at 6.

23. Field exposures not in layers

Field exposures in large, distinguishable flows . . . . . . Go to 25

Field exposures not in distinguishable flows . . . . Go to 54

24. Rock does not split along bedding plane

Rock appears to be in flows . . . . Go to 25

Rock does not appear to be in flows . . . . See hornfels, table 2

25. Rock appears to be in flows

Chip of rock held to light has translucent edges . . . See andesite, table 6

Chip of rock held to light is opaque . . . . See basalt, table 6

26. Rock will not "fizz" if acid is placed on it

Rock has greasy look; may be greenish or brown; may split along thin planes . . . . See serpentine, table 2

Rock does not have greasy look . . . . Go to 27

27. Rock does not have greasy look

Rock splits easily in thin layers . . . . Go to 28

Rock does not split easily in thin layers . . . . Go to 29

28. Rock splits easily in thin layers

Rock has shiny smooth surface . . . . See slate, table 2

Rock does not have shiny smooth surface . . . . Go to 29

29. Rock does not split easily into thin layers

Rock is red . . . . Go to 58

Rock is not red . . . . Go to 31

30. Rock does not have shiny smooth surface

Rock is red . . . . Go to 58

Rock is not red . . . . Go to 31

31. Rock is not red

Rock is soft, punky, and will burn smokily
in fire . . . . See lignite, table 5

Rock is not soft and punky and will not burn . . . . . Go to 32

32. Rock will not burn

Rock will become plastic and sticky if wet . . . . See clay, table 5

Rock will not become plastic and sticky if wet . . . . Go to 52

33. Rock may be in layers, may appear sugary

Rock tough; rings when struck . . . . Go to 34

Rock does not ring when struck . . . . Go to 43

To "ring" rock, strike it with a hammer while you are holding the rock freely in your hand; otherwise, the sound may be damped.

34. Rock tough; rings when struck

Rock breaks across grains . . . . See quartzite, table 2

No grains are distinguishable . . . . See rhyolite, table 6

35. Rock is not dark brown or black on fresh surfaces

Rock consists of rock fragments or mineral grains
of approximately the same size . . . . Go to 36

Rock does not consist of rock fragments or mineral
grains of approximately the same size . . . . Go to 37

See note at 9.

36. Rock consists of rock fragments or mineral grains
of approximately the same size

Rock may be in layers, may appear sugary . . . . Go to 33

Rock not in distinguishable layers . . . . Go to 42

37. Rock does not consist of rock fragments or mineral grains of approximately the same size

Rock consists of rounded grains and pebbles . . . See conglomerate, table 5

Rock consists of angular grains, blebs, or fragments . . . . Go to 38

38. Rock consists of angular grains, blebs, or fragments

Rock consists of angular fragments of rocks or minerals and rocks . . . . See breccia, in list of rock names and conglomerate, table 5

Rock consists of fragments of minerals only . . . . Go to 56

39. Rock is not wholly dark brown or black on fresh surfaces

Field exposures are visible . . . . Go to 40

Field exposures are not visible . . . . Go to 41

40. Field exposures are visible

Field exposures may be in layers, may be sugary . . . . . Go to 33

Field exposures not in distinguishable layers . . . . Go to 44

41. Field exposures not visible

Rock has greasy look; may be greenish or brown; may split along thin planes . . . . See serpentine, table 2

Rock does not have greasy look . . . . Go to 59

42. Field exposures not in distinguishable layers

Rock "rings" when struck . . . . See rhyolite, table 6

Rock does not ring when struck . . . . See tuff, table 6

See note at 33.

43. Rock does not ring when struck

Rock breaks easily around grains . . . . See sandstone, table 5

Rock does not break easily around grains or no grains are distinguishable . . . . See rhyolite and tuff, table 6

44. Field exposures not in distinguishable layers

Rock has greasy look; may be greenish or brown; may split along thin planes . . . . See serpentine, table 2

Rock does not have greasy look . . . . Go to 59

45. Rock consists of mineral grains or rock fragments of approximately the same size

Rock consists of broken fragments or rounded grains cemented together . . . . See sandstone, table 5

Rock does not consist of broken fragments or rounded grains cemented together . . . . Go to 48

The difference required here is between a rock that is cemented together and one in which the grains are intergrown. If the grains are separate, yet the rock adheres together, choose the first alternative; if you can see interfingering of grains with your hand lens, choose the second. Experience in studying a known sample of sandstone and one of granite will help you with this distinction.

46. Rock has some large grains, others small

Rock has wavy, banded, knotted, clotted, or gnarled appearance . . . . Go to 47

Rock does not have wavy, banded, knotted, clotted, or gnarled appearance . . . . See porphyry, table 6

47. Rock has wavy, banded, clotted, knotted, or gnarled appearance

Rock breaks along distinct lines, has flaky appearance (usually due to mica flakes) . . . . See schist, table 2

Rock does not always break evenly along distinct lines, but shows streaking of mineral grains, especially quartz, feldspar, and dark minerals . . . See gneiss, table 2; compare with breccia, under "conglomerate" in table 5

48. Rock does not consist of broken fragments cemented together

All rock grains appear to be glassy . . . . See rhyolite, table 6

Rock grains are not all glassy . . . . Go to 49

49. Rock grains are not all glassy

Rock has considerable quartz, also feldspar; may have dark minerals. Usually has a salt-and-pepper appearance . . . . See granite, table 3

Rock has very little or no quartz . . . . Go to 50

50. Rock has very little or no quartz

But has feldspar and dark minerals . . . . Go to 51

Has no feldspar, all dark minerals . . . . See peridotite, table 3

51. Rock has feldspar and dark minerals

Rock has less than 50% dark minerals . . . . See diorite, table 3

Rock has more than 50% dark minerals . . . . See gabbro, table 3

52. Rock will not become plastic if wet

Rock rings when struck with hammer . . . . Compare slate and hornfels, table 2, and basalt and andesite, table 6

Rock does not ring when struck with hammer . . . . Go to 53

See note at 34.

53. Rock does not ring when struck

Grains indistinguishable; may contain fragments of minerals, volcanic rock, or pebbles; gritty feel . . . . See tuff, table 6; compare with shale, table 5

Grains indistinguishable or barely discernible with hand lens . . . . See shale, table 5; compare with tuff, table 6

54. Field exposures not in distinguishable flows

Rock breaks in "scoops" (has conchoidal fracture). . . . . Go to 57

Rock does not break in "scoops" . . . . Go to 20

See note at 6.

55. Rock does not have "greasy" look

Rock splits along definite planes; all minerals large enough to see with hand lens . . . . See phyllite, table 2

Rock splits easily along definite planes, has distinct greenish cast . . . . See mariposite, under mica, table 1

56. Rock consists of fragments of minerals only

Rock has grains of uniform size . . . . Go to 45

Rock has some large grains, others small . . . . Go to 46

57. Rock breaks in "scoops" (has conchoidal fracture)

Rock is glassy . . . . See obsidian, table 6

Rock is not glassy . . . . See chert, table 2; compare rhyolite, table 6

58. Rock is red

Rock has spongy, porous look (but may be hard) . . . . See cinder, table 6

Rock does not have spongy, porous look . . . . See laterite, table 5

59. Rock does not have greasy look

Rock breaks in "scoops" (has conchoidal fracture). . . . See chert, table 2

Rock does not break in "scoops" . . . . Compare hornfels, table 2, rhyolite, table 6, and quartz, table 1

See note at 6.

60. Rock is vesicular

Chip of rock held to light has translucent edges . . . See andesite, table 6

Chip of rock held to light is opaque . . . . See basalt, table 6

INDEX